Synthetic Immunology

Takeshi Watanabe • Yousuke Takahama
Editors

Synthetic Immunology

Springer

Editors
Takeshi Watanabe
Kitano Hospital
Kita-ku, Osaka
Japan

Yousuke Takahama
Division of Experimental Immunology
University of Tokushima
Tokushima
Japan

ISBN 978-4-431-56025-8 ISBN 978-4-431-56027-2 (eBook)
DOI 10.1007/978-4-431-56027-2

Library of Congress Control Number: 2016939272

Printed on acid-free paper

This Springer imprint is published by Springer Nature
The registered company is Springer Japan KK

Preface for Synthetic Immunology

Immunology has developed rapidly over the last 50 years through the incorporation of new technologies and concepts in cell and molecular biology, genetics, genomics, and proteomics. This progress is the result of works by many scientists all over the world. Currently, immunological research has accumulated detailed knowledge on various mechanisms of immunity and is in the process to change medical practices. Yet, due to the enormous complexity of the immune system, many aspects on the regulation and function of the immune system remain unknown. Synthetic biology is an emerging research field that uses gain-of-function rather than loss-of-function approaches. The goals of synthetic biology can be described in a simple phrase "rebuild, alter, and understand," namely, to rebuild minimal functional systems using well-defined parts from nature and then to perturb the system to understand its working principles. Given the richness of accumulated knowledge in molecular and cellular mechanisms of the immune system, we may now begin adapting the concepts of synthetic biology to immunology. With this in mind, we have experimentally started Synthetic Immunology Workshop in 2010 and have organized four meetings, including three international meetings, by now. An immune response is a spatiotemporal phenomenon occurring at a given time and at a specialized place in the body. Thus, one goal of synthetic immunology is to reconstruct artificial microenvironments for better understanding of an immune response in four dimensions. For this reason, one central theme of synthetic immunology is "the synthesis of lymphoid niches and organs for lymphocyte development and function," which is the viewpoint of this book. We hope this yet-to-be-experimental approach of synthetic immunology and the compilation of this book will aid our further understanding of the immune system and future devising the tools to manipulate the immune system for therapy and prevention of the diseases. This book consists of the development and regeneration of immune cells (Chaps. 1 and 2), the immune organ development and artificial regeneration (Chaps. 3, 4, 5, and 6), and the synthetic approach towards understanding human immune system (Chaps. 7, 8, and 9).

Osaka, Japan Takeshi Watanabe
Tokushima, Japan Yousuke Takahama

Contents

Editors and Contributors

Editors

Yousuke Takahama, Division of Experimental Immunology, Institute for Genome Research, University of Tokushima, Tokushima, Japan

Takeshi Watanabe, The Tazuke-Kofukai Medical Research Institute, Kitano Hospital, Kita-ku, Osaka, Japan

Contributors

Kenichi Arai, Graduate School of Science and Engineering for Research, University of Toyama, Toyama, Japan

Marco Barsanti, Department of Anatomy and Developmental Biology, Monash University, Melbourne, Australia

Richard Boyd, Department of Anatomy and Developmental Biology, Monash University, Melbourne, Australia

Andrea Brendolan, Division of Experimental Oncology, IRCCS San Raffaele Scientific Institute, Milan, Italy

Ann Chidgey, Department of Anatomy and Developmental Biology, Monash University, Melbourne, Australia

Glaucia C. Furtado, Immunology Institute, Icahn School of Medicine at Mount Sinai, New York, NY, USA

Michael Hun, Department of Anatomy and Developmental Biology, Monash University, Melbourne, Australia

Tomokatsu Ikawa, Laboratory for Immune Regeneration, RIKEN Center for Integrative Medical Sciences, Yokohama, Japan

Koichi Kato, Department of Biomaterials Science, Graduate School of Biomedical Science, Hiroshima University, Minami-ku, Hiroshima, Japan

Hiroshi Kawamoto, Department of Immunology, Institute for Frontier Medical Science, Kyoto University, Kyoto, Japan

Yuka Kobayashi, The Tazuke-Kofukai Medical Research Institute/Kitano Hospital, Kita-ku, Osaka, Japan

Eric Lagasse, McGowan Institute for Regenerative Medicine, Department of Pathology, University of Pittsburgh School of Medicine, Pittsburgh, PA, USA

Elisa Lenti, Division of Experimental Oncology, IRCCS San Raffaele Scientific Institute, Milan, Italy

Sergio A. Lira, Immunology Institute, Icahn School of Medicine at Mount Sinai, New York, NY, USA

Takuya Maeda, Department of Immunology, Institute for Frontier Medical Science, Kyoto University, Kyoto, Japan

Department of Hematology and Oncology, Graduate School of Medicine, Kyoto University, Kyoto, Japan

Kyoko Masuda, Department of Immunology, Institute for Frontier Medical Science, Kyoto University, Kyoto, Japan

Takahito Mimura, Graduate School of Science and Engineering for Research, University of Toyama, Toyama, Japan

Mahmood Mohtashami, Sunnybrook Research Institute, University of Toronto, Toronto, ON, Canada

Tadashi Nakaji-Hirabayashi, Graduate School of Science and Engineering for Research, University of Toyama, Toyama, Japan

Makoto Nakamura, Graduate School of Science and Engineering for Research, University of Toyama, Toyama, Japan

Hiromitsu Nakauchi, Institute for Stem Cell Biology and Regenerative Medicine, Stanford University, Stanford, CA, USA

Division of Stem Cell Therapy, Center for Stem Cell Biology and Regenerative Medicine, The Institute of Medical Science, The University of Tokyo, Tokyo, Japan

Shreya Shukla, Institute of Biomaterials & Biomedical Engineering, University of Toronto, Toronto, ON, Canada

Jungo Tagawa, Graduate School of Science and Engineering for Research, University of Toyama, Toyama, Japan

Takeshi Watanabe, The Tazuke-Kofukai Medical Research Institute/Kitano Hospital, Kita-ku, Osaka, Japan

Daniel J. Wesche, Institute for Stem Cell Biology and Regenerative Medicine, Stanford University, Stanford, CA, USA

Hirotoshi Yoshida, Graduate School of Science and Engineering for Research, University of Toyama, Toyama, Japan

Peter Zandstra, Institute of Biomaterials and Biomedical Engineering, University of Toronto, Toronto, ON, Canada

Juan Carlos Zúñiga-Pflücker, Department of Immunology, University of Toronto, Toronto, ON, Canada

Sunnybrook Research Institute, Toronto, ON, Canada

Chapter 1
Development and Regeneration of Hematopoietic Stem Cells

Daniel J. Wesche and Hiromitsu Nakauchi

Abstract Hematopoietic stem cells (HSCs) have the capacity to maintain or reconstitute an organism's immune system for life. Due to limited supply of these rare cells, the generation of HSCs *de novo* from readily available cell sources has been a central goal in synthetic immunology. While yet unattained, progress towards that goal has been made along three major strategic avenues. First, developmental mimicking attempts to generate HSCs from pluripotent stem cells (PSCs) by recapitulating the embryonic developmental steps towards HSCs *in vitro* through temporally controlled exposure to signaling factors and inhibitors. Second, HSC development from PSCs is recapitulated randomly inside a forming teratoma *in vivo*, and hematopoietic differentiation can be enhanced and directed using hematopoietic stromal cells and cytokines. Third, combinatorial transcription factor-mediated transdifferentiation employs forced expression of transcriptional master regulators with known roles in HSCs to convert somatic cells to HSCs. In this chapter we provide a developmental perspective of HSC emergence and review the promises and challenges of the three major approaches to generate HSCs *de novo*.

D.J. Wesche (✉)
Institute for Stem Cell Biology and Regenerative Medicine, Stanford University School of Medicine, Stanford, CA, USA
e-mail: dwesche@stanford.edu

H. Nakauchi
Institute for Stem Cell Biology and Regenerative Medicine, Stanford University School of Medicine, Stanford, CA, USA

Division of Stem Cell Therapy, Center for Stem Cell Biology and Regenerative Medicine, The Institute of Medical Science, The University of Tokyo, Tokyo, Japan
e-mail: nakauchi@stanford.edu

T. Watanabe, Y. Takahama (eds.), *Synthetic Immunology*,
DOI 10.1007/978-4-431-56027-2_1

Introduction

Hematopoietic stem cells (HSCs) utilize their self-renewal and multi-lineage differentiation capabilities to supply all blood cells throughout life. As the best studied stem cell type, they have played an instructive role in the development of a number of novel ideas and models for other areas in stem cell research. Clinically, HSCs have been used in bone marrow or cord blood transplants. To prevent transplantation-associated graft-versus-host disease (GVHD), HLA matching is mandatory. This complicates further the excessive demands for genetically matched donor HSCs. Therefore; a long-standing goal in the field of hematology has been the expansion of HSCs *ex vivo*. More recently, since the advent of pluripotent stem cells (PSCs) such as embryonic (ESCs) or induced pluripotent stem cells (iPSCs), an alternative route to generating large HSC supplies has been suggested by the possibility of differentiating PSCs into HSCs.

Because of the availability of various functional assays, murine HSCs have been characterized extensively and can be isolated by fluorescence-activated cell sorting (FACS) to the level where reconstitution of the entire blood system can be achieved by transplantation of a single FACS-purified HSC (Osawa et al. 1996). The gold standard for the definition of HSCs (also long-term HSC, LT-HSC) is long-term ($>$12 weeks) multilineage reconstitution of a lethally irradiated mouse after transplantation. Several sets of (cell surface) markers have been described over the years in the mouse, all of which include c-Kit$^+$, Sca1$^+$ and lineage-marker$^-$ (lineage markers, Lin, commonly being B220 for B-cells, CD4 and CD8 for T-cells, Ter119 for erythrocytes, Gr1 for granulocytes, and IL7R for common lymphoid progenitors, CLPs), commonly abbreviated as KSL. Purification can also be achieved by using the dye efflux property of stem cells, giving rise to the side population (SP) (Goodell et al. 1996). Alternatively, Thy1.1^{low}, Flk2$^-$ significantly enrich for HSCs (Christensen and Weissman 2001). EPCR$^+$ can also assist in the purification of HSCs (Kent et al. 2009). However, emerging through the recent years as the most reliable and reproducible marker combination is KSL CD34$^-$, CD41$^-$, CD150$^+$ (Kiel et al. 2005; Morita et al. 2010). Use of these markers purifies HSCs to a degree where over 30 % of recipients show long-term engraftment in all five major blood lineages (platelets, erythrocytes, granulocytes/monocytes, T cells and B cells) after single HSC transplantation (Yamamoto et al. 2013). Considering a possibly imperfect homing efficiency and potential technical limitations such as syringe dead space, the actual HSC frequency in this population may be higher.

On the contrary, a definitive characterization of human HSCs has not been achieved yet. This is largely due to the lack of a functional equivalent to the mouse transplantation assay. Because of the xenogeneic barrier, human candidate HSCs are transplanted to immunocompromised recipients (NSG or NOG mice). The life span of these animals, even after a reconstituting transplant, is 18 months (Shultz et al. 2005), and the cell numbers required to generate a functional immune system are, due to body size, far below what would be required for a human recipient. It is thus unclear, whether what we currently believe to be the human

HSC, is in fact a true HSC that possesses the self-renewal capacity to reconstitute an entire human blood system for life at the single cell level. Accordingly, human cells that reconstitute all lineages in an immunocompromised recipient mouse have been termed SCID-repopulating cells (SRCs) rather than HSCs (Larochelle et al. 1996).

For simplicity, we will use the term HSC for SRCs as well, as it is generally assumed that the two populations largely overlap, although as mentioned, definitive proof for this is missing.

Cell populations containing less-defined hematopoietic progenitors with varying self-renewal activity and differentiation potential, or a mixture of stem and progenitor cells, such as the $CD34^+$ fraction of human CB, are commonly termed hematopoietic stem and progenitor cells (HSPCs).

Success in xenogeneic transplantation at the single cell level for human HSCs has been achieved using the marker combination Lin^- $CD34^+$, $CD38^-$, $CD45RA^+$, $CD49f^+$ (Notta et al. 2011). $Thy1^+$ and SP can further enrich HSCs in this fraction, but functional assays demonstrated the existence of Thy^- and non-SP HSCs. Interestingly, $CD34^-$ HSCs have been described as well (Anjos-Afonso et al. 2013).

Despite decades of research efforts, it has been impossible to expand or even maintain HSCs in culture *ex vivo* for more than a few passages, as they rapidly lose both their self-renewal and differentiation capacity once removed from their endogenous niches in the BM. Presumably, this is due to the lack of one or more factors that maintain and expand HSCs in their native environment *in vivo*.

For both clinical and research purposes, the availability of biological material is a major limiting factor. An allogeneic stem cell transplant for an adult patient requires about $7–10 \times 10^6$ $CD34^+$ cells per kg bodyweight (Sharp et al. 2000). The frequency of *bona fide* HSCs in the $CD34^+$ stem and progenitor compartment is about 0.5 %, with the marker combination Lin^- $CD34^+$, $CD38^-$, $CD45RA^+$, $CD49f^+$ enriching to up to 10 % of cells showing long-term multilineage engraftment upon intrafemoral injection (Notta et al. 2011).

In a research setting, the majority of assays, especially those of biochemical nature, require $10^5–10^6$ cells per sample. In our hands, one cord blood unit (~100 ml) yields about $1–3 \times 10^6$ $CD34^+$ cells.

Therefore, the ex-vivo expansion of HSCs, or the generation of HSCs *de novo* from PSCs such as ESCs or iPSCs has been a long-standing goal in hematology. Approaches, in general, follow one of three major strategies (Fig. 1.1):

(i) Mimicking of developmental steps in in-vitro culture
(ii) Directed generation *in vivo*
(iii) Transcription factor-based transdifferentiation and (re-)programming

In the following, we will describe these approaches, highlight several landmark studies, and summarize the progress made thus far.

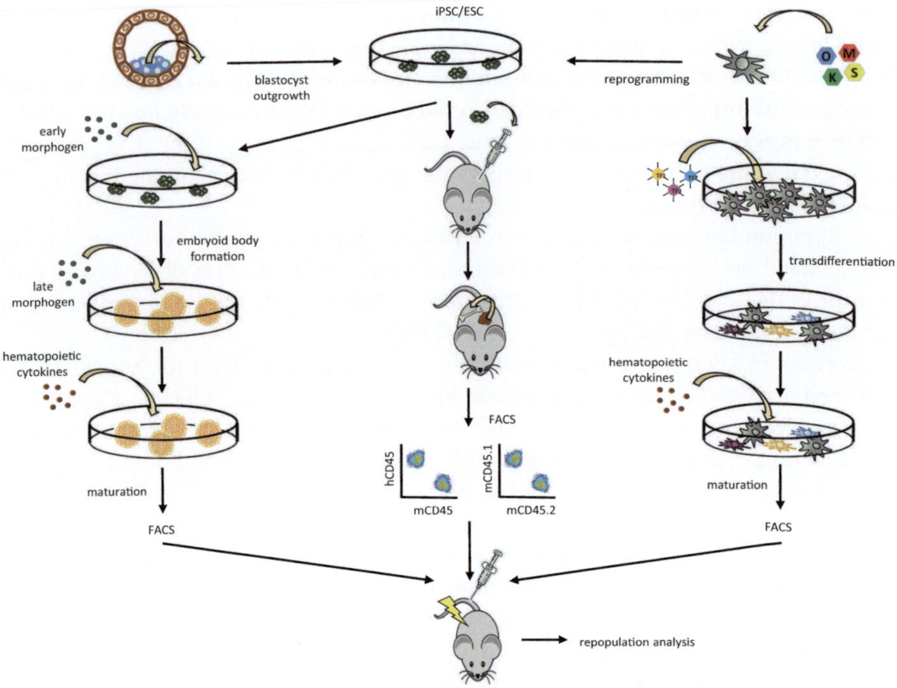

Fig. 1.1 Strategies for de-novo HSC generation. HSCs can be generated from PSCs by (1) recapitulating multi-step signaling during embryonic development in culture, or (2) by teratoma formation directed towards hematopoiesis using stromal cells and cytokines. Somatic cells can be converted into HSCs by (3) transcription factor-mediated cell fate conversion. HSC functionality is assessed by transplantation into irradiated recipients

Developmental Mimicking

In normal development, the embryo's inner cell mass (ICM) gives rise to all tissues of the organism, including HSCs. Neighboring cells and the environment provide instructive signals that, in a finely tuned spatiotemporal manner, direct differentiation of ICM cells toward specific lineages. HSC specification is achieved by a specific series of signals that, through multiple intermediate stages, differentiate a subset of the original ICM cells and their progeny into the HSC phenotype.

The most straightforward and thus most prevalent approach to the generation of various cell types from PSCs is developmental mimicking. It follows the assumption that providing the correct signals in both the right intensity and timing will induce PSCs in culture to follow the path of embryonic development toward theoretically any cell type of interest. In the case of HSCs this would mean hematopoietic differentiation to give rise to a homogenous HSC population.

Developmental mimicking has been successful to some extent, with the most advanced and well-known examples being in-vitro differentiation of PSCs into cardiomyocytes (Mummery et al. 2012) and a variety of neuronal lineages, such as neural stem cells (Chambers et al. 2009) and cortical neurons (Shi et al. 2012).

Besides from studies in human samples, our knowledge of the developmental steps involved in HSC specification largely comes from studies in mammals such as the mouse, and non-mammalian vertebrate species such as chick, zebrafish and xenopus. Blood development generally follows similar steps between mouse and human, whereas zebrafish, xenopus and chick development, while still informative, exhibits some major differences. In the following, we will discuss primarily mouse and human HSC specification, draw upon non-mammalian examples where appropriate, and highlight some instances of interspecies variability.

Embryonic Development of HSCs

During implantation in the wall of the uterus, the cells of the ICM start to lose their pluripotency and undergo their first differentiation step, as some of them generate the primitive endoderm (PrE), and the rest give rise to the epiblast. Epiblast stem cells (EpiSCs) are unable to pass stringent pluripotency tests such as tetraploid complementation, and are also called primed ESCs (as opposed to the naïve state of the cells in the ICM) (Tesar et al. 2007; Nichols and Smith 2009; Najm et al. 2011). While the transcriptomes of naïve and primed PSCs are strikingly similar, differential use of the proximal (EpiSC) versus distal (ESC) enhancers of the pluripotency transcription factor (TF) Oct4 can distinguish between the two cell types (Tesar et al. 2007; Takashima et al. 2014; Theunissen et al. 2014). The fate choice between PrE and epiblast is mediated by fibroblast growth factor (Fgf) signaling. Fgf4 expression increases in epiblast-committed cells, while PrE progenitors upregulate Fgfr2 (Guo et al. 2010).

The first event of major developmental significance to the specification of HSCs is the formation of the primitive streak (PS) from a subset of EpiSCs. PS formation is dependent on Activin/Nodal signals (Mitrani et al. 1990; Zhou et al. 1993; Conlon et al. 1994; Lu and Robertson 2004; Soares et al. 2005; Gadue et al. 2006) and Wingless-related integration site (Wnt) family proteins (Gadue et al. 2006; Sumi et al. 2008) from the posterior marginal zone (Mitrani et al. 1990). Bone morphogenetic protein 4 (BMP4), originating in the extraembryonic ectoderm (Lawson et al. 1999), while not absolutely required, enhances PS formation, specifically of the posterior PS (Nostro et al. 2008). Cells of the PS upregulate the expression of Brachyury, also known as T (Kispert and Herrmann 1994), and Mixl1 (Hart et al. 2002). Specifically the posterior PS can be identified using markers such as Evx1 (Dush and Martin 1992) and HoxB1 (Forlani et al. 2003), as opposed to anterior PS, which expresses Foxa2 (Sasaki and Hogan 1993) and Goosecoid (Kinder et al. 2001).

Following PS formation, the single layer of the epiblast is converted into the three germ layers, endoderm, mesoderm and ectoderm, by a wave of major cell migration called gastrulation (day post conception, dpc 6.0-7.5, also E6.0 in the mouse and week 2 or dpc 7-10 in humans). The PS invaginates and cells of the epiblast migrate inward and downward through the PS. The cells passing through the most anterior region of the PS form the endoderm. Cells that remain in the original position give rise to the ectoderm. The mesoderm originates from cells migrating through the more posterior portion of the PS, which then position themselves between endo- and ectoderm (reviewed in (Solnica-Krezel and Sepich 2012). Endoderm formation from the anterior PS is dependent on sustained Activin/Nodal and Wnt signaling, while BMP4 in combination with Wnt and low Activin levels induces mesodermal differentiation from the posterior PS (D'Amour et al. 2005; Sumi et al. 2008).

HSCs, like all other cells of the hematopoietic lineage, originate from the mesodermal germ layer. Hematopoiesis in the embryo occurs in three waves, a primitive, a transient definitive, and a stable definitive wave. The first mesodermal cells transiting through the posterior PS migrate to the yolk sac (YS), the site of primitive extraembryonic erythropoiesis (Kinder et al. 1999). There, they give rise to erythromyeloid-restricted progenitors that generate mainly erythroid cells, with occasional primitive-type macrophages and megakaryocytes. These progenitors cluster in YS blood islands, which are first detected at E7.0 in the mouse (Palis et al. 2001), and cease to exist at E9.0 (Palis et al. 1999). In human development, a precise description of the origins of primitive erythromyelopoiesis has been difficult due to the lack of embryos at this early stage, but blood islands have been found to arise at about dpc 16 (Luckett 1978), and the general sequence of events seems to parallel what has been described in the mouse (Tavian et al. 1999). Human primitive hematopoiesis in the YS ceases after dpc 60 (Huyhn et al. 1995). Zebrafish primitive hematopoiesis differs somewhat from its mammalian equivalent, as it is initiated inside the embryo proper (rather than the YS) from the ventral lateral mesoderm (Davidson et al. 2003).

The origin of definitive HSCs is still under debate, and several sites of origin, not mutually exclusive, have been proposed. Consensus holds, however, that lymphomyeloid progenitors are first detected after the initiation of the second (1st definitive) phase of hematopoiesis.

Within the mesoderm, a medial to lateral subdivision can be made regarding what part gives rise to which mature tissues. The medial to lateral subdivision arises from different levels of BMP4 (Tonegawa et al. 1997), which is secreted by the overlying ectoderm, and its antagonist Noggin (Tonegawa and Takahashi 1998). The temporal order of migration through the PS seems to contribute to the specification of different mesodermal lineages (Kinder et al. 2001). The lateral plate mesoderm (Foxf1^{+}, (Mahlapuu et al. 2001)) arises most laterally, and subsequently splits into two layers, the somatopleure and the splanchnopleure, which are separated by the embryonic coelom. This separation depends on BMP family member signaling, specifically BMP2 and BMP7 (Funayama et al. 1999). All definitive hematopoietic lineages are descendants of the lateral plate mesoderm, specifically the ventral part (splanchnopleuric mesoderm) (Cumano et al. 1996) that is closely associated with the endodermal layer and subsequently attaches to it to form the splanchnopleuric mesenchyme, also Foxf1^{+}. In both mouse and human, early hemogenic mesoderm can be subdivided into cardiogenic and hematopoietic mesoderm by PDGFRa expression (Liu et al. 2012).

Ventrally and medially of the splanchnopleuric mesoderm, the dorsal aorta develops by dpc 8.0 (mouse) in part from somite- (Ambler et al. 2001) and in part from splanchnopleuric mesoderm-derived endothelial cells (Sato 2013). The region of the splanchnopleure adjacent to the dorsal aorta is called para-aortic splanchnopleure (PAS) (Cumano et al. 1996), which is the precursor of the aorta-gonad-mesonephros (AGM) region, the source of the earliest intraembryonic definitive hematopoietic potential (Medvinsky et al. 1993). The AGM is where the hemangioblast develops, a multipotent progenitor cell type that can give rise to both vascular endothelial and blood lineage cells (Choi et al. 1998). Hemangioblast differentiation versus other lateral plate mesodermal lineages may depend on the absence of BMP4 signals (Gupta et al. 2006), as has been shown in zebrafish. BMP4 at later stages in mouse embryos however augments hematopoietic differentiation (Marshall et al. 2007; Durand et al. 2007), which may reflect interspecies variation or differential requirements along a temporal axis. Survival and differentiation of hemangioblasts into both hematopoietic cells and endothelial cells depends on VEGF and is independent of BMP or Activin signaling. Specifically the formation of primitive erythropoiesis, but not that of definitive hematopoietic progenitors requires simultaneous Wnt (Wnt3a) signaling (Nostro et al. 2008). Hemangioblasts are endothelial CD31^{+} Flk1^{+} VE-Cadherin^{+} CD41^{-} CD45^{-} cells and constitute the first cell type during embryonic hematopoietic development that can be isolated in sufficient purity solely based on cell surface marker expression. Isolation of earlier intermediates usually relies on the use of genetic reporters, usually expressed under lineage-specific TF promoters.

In the AGM, hemangioblasts give rise to a specific type of endothelium that harbors the potential for definitive hematopoiesis, called the hemogenic endothelium (HE) (Lancrin et al. 2009). Definitive HSCs bud off the ventral wall of the dorsal aorta (Bertrand et al. 2010; Boisset et al. 2010; Kissa and Herbomel 2010) in a process called endothelial-to-hematopoietic transition (EHT), which is dependent on Runx1 expression (Chen et al. 2009) and may be driven by BMP4 signals arising from cell clusters underlying the aortic HE (Marshall et al. 2000). The EHT imaging studies mentioned above have presented conflicting around whether the HE to HSC transition represents an asymmetric cell division, or whether the transition takes place independent of cell division with the HSC subsequently detaching from the endothelium (reviewed in (Zape and Zovein 2011)). Acquisition of CD41 expression has been suggested to mark the transition from HE to early HSCs (Mikkola et al. 2003; Ferkowicz et al. 2003; Li et al. 2003). Interestingly, these cells are negative or low for the panleukocytic marker CD45, which is only acquired by HSCs at later stages of development (Bertrand et al. 2005). The AGM-derived HSCs are the first cells capable of substantial (10–70 %) long-term multilineage reconstitution in an irradiated adult recipient (North et al. 2002). Hematopoietic cells derived from the mesenchyme underlying the ventral aspect of the dorsal aorta, or distinct subaortic patches have definitive HSC activity, but only generate low levels of engraftment in irradiated adult hosts (Bertrand et al. 2005).

In human embryos, definitive hematopoietic clusters with lymphoid potential arise in the AGM at dpc 24-27 (Tavian et al. 1999, 2001). The human immunophenotype of these earliest HSCs has been defined as $CD34^+$ VE-$cadherin^+$ $CD45^+$ c-Kit^+ $Thy1^+$ $Endoglin^+$ $CD38^{-/low}$ $CD45RA^-$ (Ivanovs et al. 2014). These cells were shown to be confined to the aortic floor and are capable of long-term multilineage engraftment in bulk. Another marker that has been proposed to distinguish between non-hemogenic mesoderm and hemangioblast/HE/early HSCs is angiotensin-converting enzyme (ACE, or CD143) (Jokubaitis et al. 2008). Use of this marker in culture however warrants caution, as it has been found to be expressed by both primitive and definitive HSCs and their precursors.

Several reports have also implicated the pre-circulation YS with some extent of definitive hematopoiesis (Yoder et al. 1997). However, these HSCs only potently engrafted newborn mice, not adults, suggesting that these early definitive HSCs may require additional maturation steps. Whether they are derived from the same lineage as the presumably mature dpc 10.5 AGM definitive HSCs is unclear, as circulation connects these regions after dpc 8 (dpc 21 in human). In addition, a studies in both mouse and human pre-circulation embryos did not find evidence for lymphoid potential in YS-derived progenitors (Müller et al. 1994; Tavian et al. 2001).

Besides the AGM and YS, the placenta has been suggested to be an independent source of HSC generation and expansion. Initial reports showed that definitive HSCs appear in the mouse placenta around the same time as in the AGM (E10.5-11.0), preceding definitive YS, FL or circulating HSCs, and expand there until E11.5-12.5 (Gekas et al. 2005; Ottersbach and Dzierzak 2005). They were found to be capable of long-term multilineage reconstitution and displayed an immunophenotype similar to AGM- and FL-HSCs (c-Kit^+ $Sca1^+$ $CD31^-$ $CD34^+$). CD41 expression surprisingly was not found on these cells, raising the question whether CD41 truly marks HSCs or is only acquired shortly after EHT. The presence of HSCs in the placenta has been confirmed in human embryos at dpc 21-42, with decline after dpc 50 (Robin et al. 2009).

The question whether HSCs are generated *de novo* in the placenta, or whether the placenta is colonized by AGM or YS-derived HSCs and progenitors, and serves as a potent niche for HSC expansion, remains unresolved. As blood flow to the placenta traverses the dorsal aorta, colonization is a likely scenario. However, pre-circulation placental tissue was found to give rise to hematopoietic colonies after explant culture (Zeigler et al. 2006; Corbel et al. 2007). Similarly, $Ncx1^{-/-}$ mice, which lack circulation, generate hematopoietic progenitors in the placenta (Rhodes et al. 2008). E8.5 lethality however precludes transplantation experiments in this setting. Thus, indirect evidence suggests the placenta as an independent source of HSCs, but definitive proof for its independence rather than its colonization has not yet been provided. In addition, HSCs have been found in the umbilical and vitelline arteries at time points similar to the generation of HSCs in the AGM (de Bruijn et al. 2000).

Taken together, these findings indicate that HSCs are generated independently in both the AGM and the YS. While AGM-HSCs are definitive, YS-HSCs are primitive, but capable of definitive maturation upon receiving the proper environmental signals. Definitive HSCs may also arise independently in the placenta and umbilical/vitelline arteries, or these sites may be colonized early on by AGM-HSCs.

Soon after the specification of definitive HSCs in the AGM, placenta, and umbilical/vitelline arteries around dpc 11 in the mouse (Müller et al. 1994), these cells migrate to and colonize the fetal liver (FL) (Le Douarin et al. 1984). In the FL, a massive expansion of HSCs takes place (Ema and Nakauchi 2000), likely a combination of ongoing migration from other tissues and independent expansion in the FL (Takeuchi et al. 2002; Kumaravelu et al. 2002). From dpc 12 to birth, hematopoiesis takes place in the liver, representing the second wave of hematopoiesis. HSC activity in the FL decreases after dpc 16 (Ema and Nakauchi 2000), but FL hematopoiesis continues until early postnatal life.

From the FL, HSCs seed to the spleen and the bone marrow (BM). In the BM, the site of the third wave of hematopoiesis which continues throughout adult life, first definitive HSC activity can be detected at dpc 17.5 (Christensen et al. 2004). Production of peripheral blood however continues in the FL and is not taken over by the BM until after birth (Blazsek et al. 2000), at which point the BM sustains hematopoiesis throughout adult life. Not until the second week after birth, however, acquire HSCs their adult phenotype (Kikuchi and Kondo 2006).

In human, the AGM ceases HSC production after dpc 40 (Oberlin et al. 2002). BM colonization takes place around week 10.5.

From a regenerative standpoint, both FL- and BM-HSCs are more or less equivalent in their therapeutic potential, as they both are able to regenerate all mature lineages upon transplantation. In fact, FL-HSCs, which are abundant in the circulation around birth and can be isolated from cord blood, are more efficient at regenerating an irradiated recipient's immune system than BM-HSCs, both in mice (Morrison et al. 1996) and men (Kim et al. 1999). For any in-vitro system trying to recapitulate development of HSCs, the postnatal switch from fetal to adult hematopoiesis would thus likely not be required, as it would presumably occur post-transplant in the recipient's BM. In how far AGM-HSCs are equivalent to FL-HSCs in the sense that they are capable of long-term multilineage reconstitution in the setting of an irradiated recipient, remains to be tested at the most stringent, single-cell level. Regardless of the source of the HSC (AGM vs. YS vs. placenta vs. vitelline/umbilical arteries), in stark contrast to the signaling events required for specification of the HE, the events after this stage are far less well characterized.

The role of Notch1 in EHT has been fairly well documented. While leaving YS HSPCs and HE intact, definitive hematopoiesis in the AGM was found severely impaired upon Notch1 deletion (Kumano et al. 2003). Consistent with this notion, AGM hematopoiesis is markedly decreased upon deletion of Notch1 pathway components (Yoon et al. 2008) or targets (Robert-Moreno et al. 2005). In zebrafish, it has been suggested that the role of Notch1 is the activation of the EHT TF Runx1 (Nakagawa et al. 2006) via Gata2 (Robert-Moreno et al. 2005) and Ets or Scl

(Nottingham et al. 2007), though a more general, Runx1-independent role for Notch1 in EHT is possible, similar to its role in ESCs (Meier-Stiegen et al. 2010).

The involvement of Wnt signals in hematopoiesis has mostly been documented in later stages of development and adult life. As such, Wnt appears to serve as a positive modulator of HSC self-renewal rather than emergence. More recently, catecholamines secreted by cells of the sympathetic nervous system beneath the dorsal wall of the aorta have been implicated with supporting EHT in the AGM support (Fitch et al. 2012).

Unfortunately, the vast array of hematopoietic growth factors and cytokines commonly used in culture *in vitro* (SCF, TPO, Flt3-L, IL-3, IL-6, and others) has only been investigated in culture systems or after the specification of definitive HSCs *in vivo* (reviewed in (Robin and Durand 2010)). Whether these factors have a role in the specification of HSCs in the AGM and other regions, or whether they mainly serve to maintain and expand HSCs after their initial generation is currently unresolved.

Pluripotent Stem Cell Differentiation Along Developmental Pathways

Based on our knowledge of the signaling events that orchestrate lineage choices along the pathway of hematopoietic differentiation during embryonic development, various distinct cell states have been generated from PSCs *in vitro*. Three fundamental approaches have been used to promote differentiation of PSCs: i. aggregation of ESCs into three-dimensional structures known as embryoid bodies (EBs) that partly recapitulate differentiation in early embryos, ii. culture of ESCs on a layer of supportive stromal cells providing the desired environmental cues, and iii. ESC culture as a monolayer on extracellular matrix proteins to provide adhesive support. Lineage-specific differentiation is generally achieved by the addition of signaling agonists (growth factors, cytokines, small molecule agonists) or inhibitors to the culture medium. All three approaches have their advantages and disadvantages, and differentiation to lineages of interest has been achieved in either system. In fact, a substantial fraction of our knowledge on the cues required for early embryonic fate decisions comes from in-vitro culture, as mapping signals in the in-vivo environment of the developing embryo, due to the complexity of interactions and the number of concurrent signals, is increasingly difficult as development progresses.

Studies attempting to differentiate murine PSCs into hematopoietic lineages were initiated after the spontaneous appearance of hematopoietic cells in EBs had been noted (Evans and Kaufman 1981; Doetschman et al. 1985). When differentiation media were supplemented with methylcellulose to support hematopoietic development (Wiles and Keller 1991), these hematopoietic cells contained a minor fraction of Flk1^{+} progenitors, likely hemangioblast-like cells, that were capable of engrafting sublethally irradiated mice for up to 28 days, with long-term engraftment not being assessed (Miyagi et al. 2002). As one would expect, endothelial contribution was noted as well.

Subsequent studies of hematopoietic differentiation *in vitro* sought to enhance hematopoietic development in EBs. These studies used supplementation of differentiation media with hematopoietic cytokines such as SCF and IL-11 (Keller et al. 1993). In addition, EBs could be dissociated after 2–5 days of differentiating culture (Kennedy et al. 1997) and re-plated in methylcellulose medium supplemented with SCF, VEGF and other cytokines (Choi et al. 1998), or medium conditioned by endothelial cell lines. Under these conditions, blast colonies with endothelial and hematopoietic potential (counted in colony-forming units, CFU) were formed that were later shown to arise from hemangioblast-like cells (Lancrin et al. 2009).

Attempts to generate more primitive precursors were more successful in different conditions. EB culture yielded HSPCs with multilineage reconstitution ability and potentially limited self-renewal (up to 12 weeks) when $CD45^+$ (Hole et al. 1996) or $CD45^+$ $c\text{-}Kit^+$ (Burt et al. 2004) cells were isolated from EBs cultured with SCF, IL-3 and IL-6. These studies, however, have suffered from poor reproducibility, suggesting that minor differences in the individual culture conditions may be of great importance under such ill-defined conditions. In addition to EB and stromal co-cultures, hematopoietic differentiation was shown to occur in monolayers when grown on collagen IV support (Nishikawa et al. 1998).

Soon after the generation of human ESC lines in 1998 (Thomson et al. 1998), the differentiation strategies developed in mice were applied to hESCs in order to generate hematopoietic progenitors and stem cells. The earliest approaches (Kaufman et al. 2001) cultured hESCs on irradiated feeder cells that had been derived from stromal cells from the mouse BM (S17) or YS (C166), and that had been shown to support expansion of murine hematopoietic progenitors before (Wineman et al. 1996; Lu et al. 1996). While expressing CD34 and in part CD31, hematopoietic cells derived under these conditions lacked surface expression of CD45. Differentiation in this system was induced by the addition of serum, which readily and non-specifically differentiates ESCs. The hematopoietic yield could be increased by the addition of posterior mesoderm-inducting BMPs (BMP2, -4 and -7) early in culture, and hematopoietic cytokines (SCF, IL-3, IL-6, VEGF, G-CSF, Flt3-L and Epo) at d6 of differentiation (Lu et al. 2004). $CD34^+$ $CD38^-$ cells isolated from these cultures showed expression patterns similar to BM-derived progenitors. However, engraftment potential was not analyzed. Soon after, the OP9 stromal cell line was shown to be more efficient than S17 or MS-5 (Mitjavila et al. 1998) at hematopoietic induction (Vodyanik et al. 2005), and OP9 is since the predominant stroma used for in-vitro differentiation today. CD45 expression appeared soon after CD34 and CD31, suggesting that OP9-secreted factors are more efficient in hematopoietic specification than S17 and other stromal lines, which predominantly seem to induce a bipotent, hemangioblast/HE-like state marked by CD31, CD34, Flk-1 and VE-Cadherin (Wang et al. 2004). CD41 acquisition was further found to mark definitive HSCs derived in culture

(Matsumoto et al. 2009), and also to distinguish embryonic HSCs from adult HSCs. The use of OP9 lines that express high levels of exogenous morphogens required for hematopoietic specification can further enhance the beneficial effects of OP9 co-culture. This has been demonstrated for OP9-Wnt3a (Woll et al. 2008) or OP9-DL1 (Schmitt et al. 2004), the latter (DL1, delta-like 1) being an important Notch ligand.

Similar to murine cultures, human hematopoietic CFUs can be generated from EBs. Hematopoietic progenitors were increased upon addition of BMP4 and hematopoietic cytokines (Chadwick et al. 2003). An analysis of the temporal progression of hematopoietic differentiation in EBs cultured in methylcellulose medium confirmed the similarities to embryonic blood island development in the YS (Zambidis et al. 2005). As in stromal co-culture experiments, CD34 and CD31 appeared around d3, about a week prior to CD45 (after d10), and peaked at d12-15 of differentiation. Mesodermal HE-like colonies could first be generated after d7. Similar to embryonic development, these colonies initially display an endothelial phenotype and initiate budding of hematopoietic blasts after 2–5 weeks. Differentiated progeny of these colonies revealed primarily primitive macrophages and primitive nucleated erythroblasts. Definitive hematopoietic colonies were only detected in EBs cultured for 2–3 weeks, consistent with the appearance of AGM-derived definitive HSCs multiple days after YS-derived primitive progenitors.

Notably, while required for the EHT of the hemangioblast, VEGF treatment skewed hematopoietic differentiation of EBs to erythrocyte progenitors (Cerdan et al. 2004).

Culture of EB-derived $CD31^{+}$ $CD34^{+}$ $CD45^{-}$ $Flk1^{+}$ $VE\text{-}Cadherin^{+}$ hemangioblast-like cells in HSPC-expansion medium efficiently induces hematopoietic differentiation at the expense of endothelial potential (Wang et al. 2004), associated with an increase in CFU potential. Direct intrafemoral injection of hemangioblastic progenitors into NOD/SCID mice yielded low engraftment up to 8 weeks post-transplant (Wang et al. 2005). Conversely, culture in endothelial growth medium containing pituitary extracts and VEGF-A promotes endothelial differentiation at the expense of hematopoietic commitment.

After the insufficiency of either approach to generate functional equivalents to the definitive long-term HSC had been demonstrated, subsequent attempts sought to combine the efficient replication of early embryonic development in EBs with the differentiation and maturation benefits of stromal co-culture. Following culture of EBs in mesodermal/hematopoietic specification medium containing among other factors BMP4, TGFb1, low-dose VEGF, TPO, Flt3L and SCF, EBs were dissociated after 7 days and plated on OP9 stromal feeders in identical medium. Following culture for additional 7 days, cells were then re-plated in HSPC expansion medium (SCF, TPO, Flt3L, Epo, PGE2) and subjected to colony-formation assays or transplanted into NSG mice after additional 7 days (Woods et al. 2011). While these conditions led to the isolation of HSPCs phenotypically similar to CB-derived HSPCs ($CD45^{+}$ $CD34^{+}$ $CD38^{-}$), that stably formed multilineage hematopoietic colonies, engraftment in NSG mice was low (2/38 mice, peak 2.1 % chimerism after

10 weeks), suggesting that these conditions either failed to generate *bona fide* HSCs, or the generated HSCs had differentiated by the time of transplantation. Since it is unclear what distinguishes the development of primitive HSCs in the YS from the definitive HSCs arising in the AGM, it is conceivable that most developmental mimicking approaches generated YS-like rather than AMG-type HSCs. This is supported by the notion that most PSC-derived hematopoietic precursors require long maturation culture on stromal support, paralleling the notion that primitive YS and PAS HSPCs could be converted to LT-HSCs by co-culture with AGM stromal cells (Matsuoka et al. 2001). On the contrary, AGM-HSCs require no further maturation *ex vivo*, and are capable of long-term multilineage reconstitution directly upon isolation. Accordingly, deciphering the differences between HSC specification in the YS vs. the AGM may aid derivation of *bona fide* HSCs from PSCs *in vitro*.

Concurrently, protocols for hematopoietic induction from PSCs have benefited from a more narrow reproduction of the early events required for ultimate hematopoietic conversion. In the presence of Nodal/Activin and low serum, ESC differentiation in monolayer culture led to the induction of a PS-like population after 36 h (D'Amour et al. 2005). Using a Mixl1-GFP reporter, early PS cells could be isolated from EBs differentiated in the presence of Activin A and BMP4, but not Fgf2 (Davis et al. 2008). Similar data has been presented for BMP4 (Park et al. 2004), but not Fgf2, VEGF and SCF as inducers of PS/mesoderm formation (Pick et al. 2007). EB differentiation however generates a variety of morphogens and growth factors in unknown concentrations. The system therefore is capable of identifying necessity, but it is insufficient to demonstrate sufficiency of any morphogen. In a more stringent approach, using a CD4-Foxa2 reporter, Gadue and colleagues investigated the requirements for Wnt and Nodal signaling by using selective inhibitors (Gadue et al. 2006). Complete inhibition of either pathway abolished formation of both anterior and posterior PS populations. Low-level inhibition of Wnt favored the emergence of anterior PS populations, while low-level Nodal inhibition promoted posterior PS formation. Accordingly, continuously high Nodal/Activin signaling drove subsequent differentiation to endodermal fates, while low levels promoted mesoderm formation. Hematopoietic potential was most prevalent in high concentrations of Wnt. In a follow-up study, researchers around Gadue corroborated evidence for a Wnt/Nodal requirement, even in the presence of BMP4 (Nostro et al. 2008). In addition, while BMP4 addition drove posterior PS marker expression, both anterior and posterior PS populations formed under BMP4 inhibition in Wnt/Activin-induced cultures. Similarly, the transition from T^+/$Flk1^-$ PS to T^+/$Flk1^+$ hematopoietic mesoderm required only the addition of exogenous Activin, not Wnt or BMP4. However, addition of Wnt3a and BMP4 enhanced the subsequent VEGF-driven transition to $CD41^+$ hematopoietic cells, and inhibition of either pathway severely attenuated Flk1 upregulation and inhibited CD41 acquisition. Interestingly, Flk1+ HE only required VEGF for the transition to hematopoietic cells. Short-term exposure to VEGF was sufficient for CD41 acquisition and CFU potential in methylcellulose, while longer treatment augmented erythroid differentiation. Inhibition of Nodal/Activin, Bmp and most significantly Wnt pathways inhibited primitive erythroid differentiation, but had no significant effect on the formation of definitive hematopoietic precursors. In addition, non-canonical Wnt signaling pathways have been reported to contribute to early steps of mesoderm specification in EB differentiation, thereby enhancing blood lineage commitment (Vijayaragavan et al. 2009). Whether this is required *in vivo* is unclear.

In summary, based on both analyses of natural embryonic development and the in-vitro differentiation studies described above, a general four-step method for the generation of hematopoietic progenitors from PSCs has emerged (Fig. 1.2). First, the formation of PS-like cells is induced by the simultaneous activation of Wnt/β-Catenin and Activin/Nodal signaling. PS cells are marked by the expression of T or Mixl1. Addition of BMP4 directs differentiation toward a more posterior PS fate (Foxa2$^{low/-}$) and enhances subsequent mesoderm differentiation. Second, the action of low levels of Activin combined with Wnt3a and BMP4 signaling further differentiates posterior PS cells to Flk-1^{+} T^{+} VE-Cadherin^{+} hematopoietic mesoderm/hemangioblast. Third, VEGF and bFGF induce Flk-1^{+} VE-Cadherin^{+} CD34^{+} CD31^{+} CD41^{-} CD43^{-} HE specification from hematopoietic mesoderm, and, potentially in cooperation with BMP4, stimulate EHT to yield CD41^{+} hematopoietic cells. Lastly, maturation of CD34^{+} primitive precursors into CD45^{+} hematopoietic cells is achieved by the addition of hematopoietic cytokines and growth factors, including SCF, TPO, Flt3L, IL-3 and IL-6. Additional undefined factors produced from stromal feeder or FL cell lines can be used to enhance the maturation process.

HSC Generation in vivo

As we discussed in the previous section, HSC development relies on many intricate interactions of signaling molecule expression. Despite considerable advances in our understanding of the signals that direct lineage specification, a precise spatiotemporal description of signals throughout all stages of definitive HSC specification is beyond our current technical capabilities. Especially late events in hematopoiesis and HSC specification, i.e. after the stage of HE, are only described rudimentarily. While the use of hematopoietic cytokines, co-culture with stromal cells, or a combination of the two have had some moderate successes in differentiating PSCs *in vitro* to hematopoietic fates, all in-vitro derived HSPCs exhibit functional differences when compared to native HSCs. The likely explanation lies in our lack of understanding of the precise signals that are required for the generation and maintenance of HSCs that are readily provided by the respective microenvironments *in vivo*. It is essential to establish a culture system that can maintain HSCs *in vitro*.

A sensible approach to circumventing the problem of the microenvironment complexity is to use the in-vivo microenvironment to direct the PSCs of interest towards the hematopoietic lineage. For example, when injected into an early blastocyst that is then transferred into a surrogate mother and allowed to develop

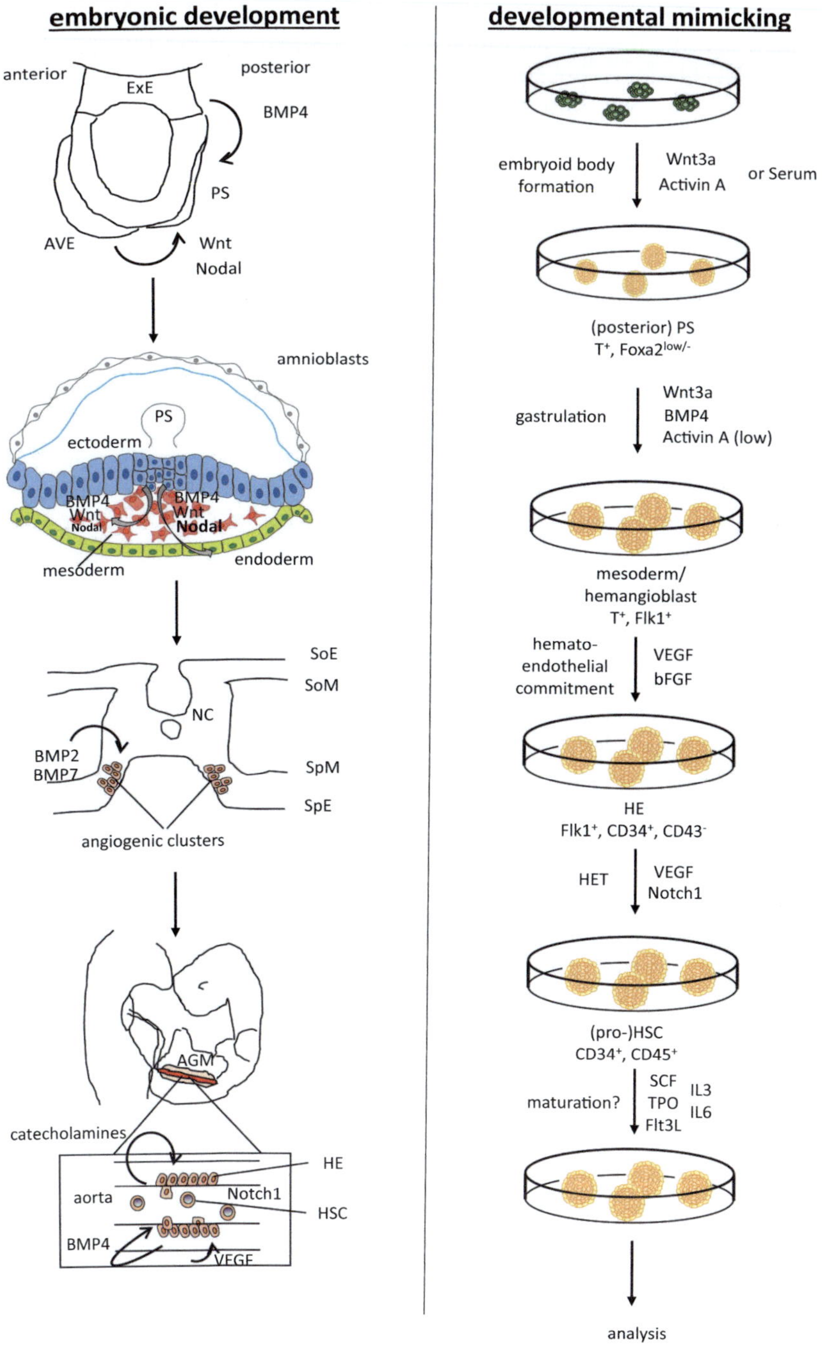

Fig. 1.2 Developmental mimicking. Left panel: embryonic development of HSCs. PS formation is induced by BMP4 from the ExE and Wnt/Nodal from the AVE. Upon gastrulation, BMP4 and Wnt combined with low Nodal signaling promote mesoderm differentiation from the posterior PS, while high Nodal favors ectoderm. Angiogenic clusters arise from the splanchnopleuric mesoderm

to term, mouse PSCs contribute to all somatic lineages, including HSCs. Contribution to one specific lineage, for example HSCs, can be enhanced by using genetically modified blastocysts that are deficient for the lineage of interest. In this case, the wild type donor PSCs fill the developmental niche and all mature cells of this lineage will be donor-derived (blastocyst complementation) (Chen et al. 1993) Surprisingly, blastocyst complementation has been shown to be successful despite a xenogeneic barrier (Kobayashi et al. 2010).

However, especially with the goal of generating human HSCs, ethical concerns and technical limitations currently preclude the use of these techniques. Ethical concerns regarding contribution of human PSCs to organs other than the intended lineage, especially the brain, and the resulting humanization of the host animal. As far as technology is concerned, neither has contribution of human PSCs to host animals been shown yet, nor do we have large animal models in which HSC development is precluded genetically.

Besides blastocyst injection or complementation, only one experimental system more of less faithfully recapitulates embryonic development. The only artificial system that generates cell types of all mature tissues is teratoma formation. A teratoma is a PSC-derived benign tumor that contains mature derivatives of all three germ layers (Evans and Kaufman 1981; Gertow et al. 2004; Buta et al. 2013). A vast array of mature cell types has been detected in teratomas, including, but not limited to, neurons, glial cells, chondrocytes, osteoblasts, goblet cells, intestinal epithelium, and endothelial cells. Considering the presence of both bone (including marrow-like structures) and vascular endothelium, it is likely that hematopoietic cells, including HSCs, may be generated in a teratoma.

Surprisingly, the capacity of a teratoma to generate functional HSCs has not been tested until recently (Fig. 1.3). Based on the idea that if mature HSCs are generated within a teratoma, they may follow natural homing cues and may migrate to the host BM, we carefully examined the BM of teratoma-bearing mice and found that a small but substantial number of miPSC-derived HSCs had colonized the host BM. These HSCs were morphologically and functionally equivalent to primary wild type HSCs (Suzuki et al. 2013). Co-injection of OP9 stromal cells or administration of hematopoietic cytokines increased the percentage of donor-derived BM HSCs. Similarly; human HSCs were derived from hiPSCs transplanted into

---

Fig. 1.2 (continued) that is induced by BMP2 and BMP7. In the AGM, HSCs bud off the aortic wall from HE. EHT is mediated by Notch1 and VEGF signaling, and supported by BMP4 and catecholamines. Right panel: four-step method to induce hematopoietic differentiation in human embryoid body culture (similarly applicable to monolayer culture). Wnt3a and Activin A (or serum) induce differentiation of PSCs, giving rise to PS-like T^{+}, $Foxa2^{low/-}$ cells. Wnt3a, BMP4, and low levels of Activin A then induce T^{+}, $Flk1^{+}$ hematopoietic mesoderm. VEGF and bFGF signaling subsequently assist development of HE ($Flk1^{+}$, $CD34^{+}$, $CD43^{-}$). HET is dependent on sustained VEGF signaling and can be enhanced by Notch1 activation. Maturation of hematopoietic cells is then promoted by addition of hematopoietic cytokines (SCF, TPO, Flt3L, IL-3 and IL-6) or OP9 co-culture, and assessed by transplantation

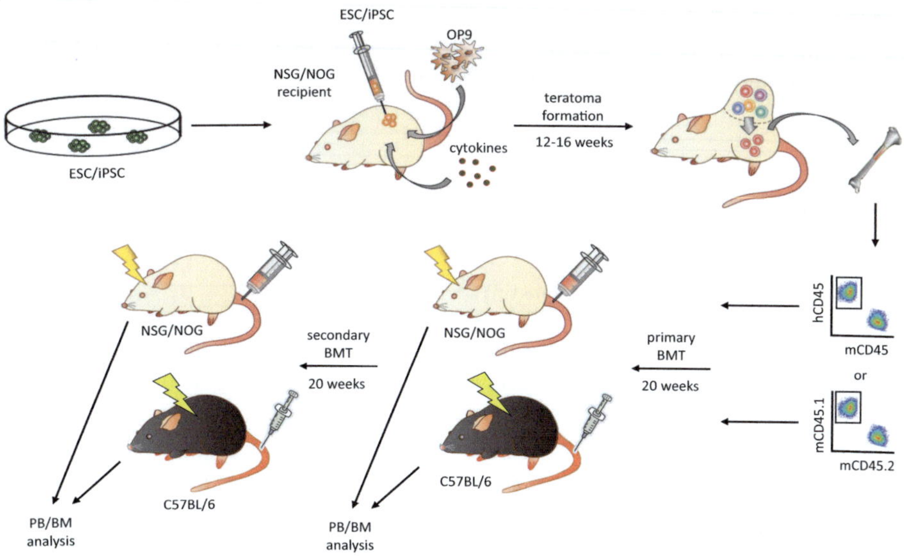

Fig. 1.3 Intrateratomal HSC generation *in vivo*. PSCs are transferred into the flank of an NSG/NOG recipient. Co-injection of OP9 stromal cells and administration of hematopoietic cytokines greatly enhances hematopoietic intrateratomal differentiation. HSCs migrate out of the teratoma and colonize the BM, which is harvested after 12–16 weeks, and donor-derived HSCs can be subsequently transplanted into irradiated (mouse) or NSG/NOG (human) recipients

NOD/SCID/JAK3^{null} mice. Upon transplantation, these HSCs showed multilineage contribution 12 weeks post transplant.

The intra-teratomal derivation of human HSCs *in vivo* was also reported by Amabile and colleagues (Amabile et al. 2013). CD45$^+$ CD34$^+$ hematopoietic cells were detected 4 and 8 weeks after PSC injection into NSG hosts. Upon co-injection with OP9-Wnt3a, hematopoiesis was markedly increased. Teratoma-derived CD34$^+$ cells consistently generated hematopoietic colonies when cultured in methylcellulose medium, and were also capable of multi-lineage reconstitution of NSG recipients. Primary recipients showed BM chimerism percentages similar to those transplanted with CD34$^+$ cells isolated from CB. While CB-derived cells mainly generated CD19$^+$ B-cells, teratoma HSCs predominantly contributed to CD15$^+$ myeloid cells, indicating some functional differences between the two cell sources. Chimerism in secondary transplantations was very low (0.04 %), suggesting limited self-renewal activity of teratoma-derived HSCs or technical limitations of xenogenic transplants.

These two studies, for the first time, demonstrated that functional HSCs can be derived from both murine and human PSCs. While their complete equivalence to native HSCs remains to be shown, especially in the setting of human teratomas, murine teratoma-derived HSCs appear to be *bona fide* HSCs. Importantly, teratoma formation constitutes the first experimental system that faithfully recapitulates the development of definitive hematopoiesis. It thus may close an important gap in our experimental toolbox and provide an attractive means for further dissection of the developmental steps involved in HSC specification from PSCs and maturation, most importantly the events required after HE emergence and EHT.

Transdifferentiation

Over the last decade, the notion has developed that cellular states are maintained by the combinatorial effect of TFs and transcriptional regulators (TRs) expressed simultaneously. While signals from the environment are important to direct lineage choices in virtually any differentiation process, they do so by activating intracellular signaling cascades that eventually result in changes of the cellular transcriptome. These changes are mostly orchestrated by TFs that act as downstream effectors of environmental signals. Cells integrate various signals to give rise to the specific TF combination required for a specific cell type.

Dependent on their current proteome, different cell types exhibit differential responsiveness to extracellular signals. In the context of hematopoietic development *in vitro*, this means that efficient conversion of PSCs to HSCs requires mimicking of every step along the natural developmental sequence, with each step relying on different signals. Failure to sufficiently recapitulate any of those intermediates may disrupt HSC potential of the final cell type. In contrast, TF networks are specific to defined cellular states (Wilson et al. 2010), tend to reinforce one another (Chew et al. 2005), and thus most likely encode all properties of the corresponding state (Rao et al. 2002; Bonzanni et al. 2013).

This notion has opened the door to circumventing the need for extracellular signals to reach defined cellular states, by the forced expression of the TFs/TRs downstream of the signal. Remarkable cellular plasticity has since been noted (Graf and Enver 2009), suggesting that the forced expression of any TF network in any unrelated cell type could reprogram the cell to the cell type maintained by the TF network used. Using stable genomic integration through viral delivery, or transient expression with episomal DNA or mRNA, many cell types have since been interconverted (Weintraub et al. 1989; Xie et al. 2004; Vierbuchen et al. 2010; Ieda et al. 2010; Sekiya and Suzuki 2011).

This culminated in the discovery of iPSCs, which can be generated from fibroblasts (Takahashi and Yamanaka 2006; Takahashi et al. 2007) and other cell types by the ectopic expression of Oct4, Sox2, Klf4 and cMyc (mouse and human), or Oct4, Sox2, Nanog and Lin28 (human) (Yu et al. 2007). Alternative factor combinations, sometimes depending on the cell type of origin, have since been described (Kim et al. 2009; Utikal et al. 2009; Giorgetti et al. 2009; Maekawa et al. 2011; Montserrat et al. 2013).

Similar approaches have been used in both mouse and human systems to generate HSCs and more restricted hematopoietic progenitors from a variety of cell types (Fig. 1.4).

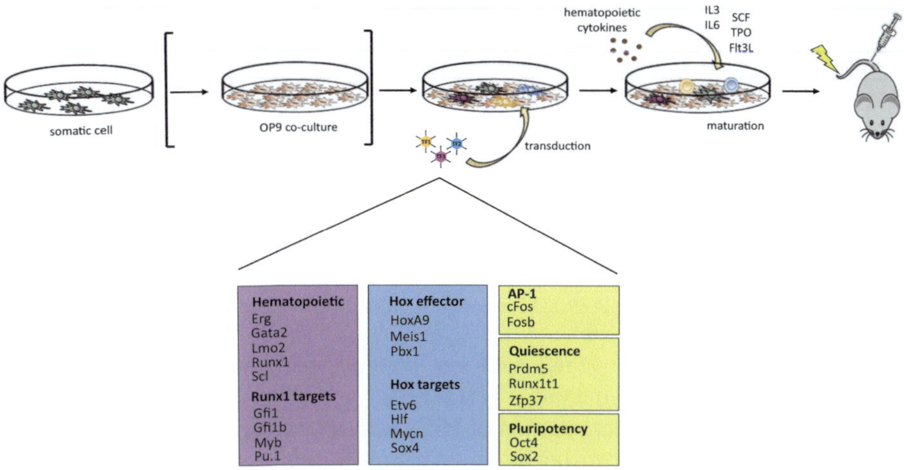

Fig. 1.4 Transcription factor-mediated transdifferentiation. Somatic cells are switched to hematopoietic cell fates by combinatorial forced expression of master transcription factors. A variety of factors has been reported to confer (some) HSC properties on cells, and the factors can be roughly divided into three groups: core hematopoietic factors and their targets, Hox effector and target genes, and regulation of cellular activity (transcriptional activation and repression). After viral transduction (or any other gene delivery method), the hematopoietic fate switch can be assisted by supplementation with hematopoietic cytokines such as SCF, TPO, Flt3L, IL-3 and IL-6, or culture on stromal support cells. The culture is analyzed by transplantation

Two studies have attempted to convert mouse fibroblasts to hematopoietic progenitors. Pereira et al. (Pereira et al. 2013) started with a cocktail of 18 hematopoietic TFs and, using viral delivery, inducibly expressed them in fibroblasts that were then plated on supportive stroma and cultured in medium supplemented with hematopoietic cytokines. They identified cFos, Etv5, Gata2 and Gfi1b as sufficient for the induction of an endothelial/hematopoietic program. However, clonogenic potential *in vitro* was only induced in these progenitors after culture on placental stroma, and engraftment *in vivo* was not assessed in this study. A second study, using constitutive expression of a set of 19 hematopoietic TFs, demonstrated a similar conversion of fibroblasts into hematopoietic progenitors via an endothelial intermediate (Batta et al. 2014). The five TFs Erg, Gata2, Lmo2, Runx1c and Scl induced an in-vitro clonogenic progenitor cell type with multilineage differentiation potential. Engraftment in vivo however was limited to short-term erythroid lineage contribution.

A recently published study by Riddell et al. for the first time reported the generation of engraftable murine HSCs using inducible TF overexpression (Riddell et al. 2014). The researchers converted committed hematopoietic precursors to HSCs competent of multilineage engraftment by screening a set of 33 TFs and three transcriptional regulators that were overrepresented in HSCs compared to committed progeny. They identified six genes, Hlf, Runx1t1, Pbx1, Lmo2, Prdm5, and Zfp37, to be sufficient for conversion, presumably aided by instructive signals in an in vivo environment. Meis1 and Mycn markedly increased the conversion rates.

Thus far, the results have not been reproduced independently.

Two studies in human fibroblasts used a slightly different approach: ectopic expression of pluripotency factors (Oct4 and Sox2, respectively) was used to attempt partial reprogramming without entering a stable pluripotent state (Szabo et al. 2010; Pulecio et al. 2014). Subsequently, hematopoietic differentiation was induced by culture in hematopoietic cytokine-containing medium. While both approaches succeeded in generating progenitor-like cells that were clonogenic *in vitro*, engraftment analysis *in vivo* revealed a low expression of CD45, indicating incomplete hematopoietic conversion.

Other attempts to induce HSCs have involved alternative source cell types that are more closely related to the target cell type.

As described above, differentiation of PSCs is fairly efficient up to the HE stage, after which differentiation to hematopoietic progenitors appears rather random, probably owing to our lack of understanding of the precise molecular pathways involved in HSC specification and maintenance. Doulatov and colleagues used $CD34^+$ $CD45^+$ myeloid-restricted hematopoietic progenitors and transduced them with a library of nine TFs that were underrepresented in this population compared to the $CD34^+$ $CD38^-$ HSPC fraction of cord blood (Doulatov et al. 2013). They detected a fraction of $CD34^+$ $CD38^-$ $CD90^+$ $CD49f^+$ HSC-like cells in progenitors transduced with Erg, HoxA9 and Rora. These cells could be endowed with in-vivo multilineage reconstitution activity upon additional transduction with Sox4 and Myb, however, reconstitution was restricted to short-term multilineage contribution.

In a similar attempt, another study aimed to induce EHT in human endothelial cells derived from the umbilical vein or dermal microvasculature (Sandler et al. 2014). Transduction with constitutive Fosb, Gfi1, Runx1 and Spi1 led to the generation of hematopoietic cells when concomitantly cultured on an endothelial cell line. These cells contained HSCs according to immunophenotype by FACS, and had activity of long-term primary and secondary multilineage engraftment *in vivo*. However, $CD3^+$ T-cell development was precluded by the constitutive nature of TF expression, and while inducible expression allowed for weak T-cell potential, the potential of HSPCs derived with the inducible system was not assayed *in vivo*.

Interestingly, for neither mouse nor human transdifferentiation, not a single TF has been identified so far to be essential for hematopoietic conversion. In contrast, TF-based approaches to induce pluripotency have identified Oct4 as a master regulator that cannot be replaced by other factors.

The ectopic expression of HoxB4 has been shown to confer definitive HSC activity on murine hematopoietic progenitors isolated from the YS or differentiated in vitro from PSCs (Kyba et al. 2002). The same strategy has been used successfully for proof-of-principle experiments in a murine model of sickle cell anemia (Hanna et al. 2007). ESC-derived hematopoietic progenitors were then converted to definitive HSCs by HoxB4 expression and were able to contribute to low-level reconstitution of gene-corrected hemoglobin. Disappointingly, HoxB4 expression failed to have the same effect on human ESC-derived progenitors (Wang et al. 2005).

Future Directions

While major advances have been made on the path to the de-novo generation of HSCs, the final goal has not been reached yet.

Two major gaps in our knowledge of HSC biology prevent further progress.

First, the events following HE specification that lead to the generation of definitive HSCs lack precise description. This is mainly due to the fact that dissection of these steps is inherently difficult. Accordingly, all current approaches to generating HSCs from HE or earlier cell types rely on a variety of factors shown to promote hematopoiesis in general, but not specifically HSC specification. Culture on stromal cell lines, most commonly OP9, provides a variety of unknown adhesion molecules and soluble factors that promote HSC maturation. Hematopoietic cytokine cocktails commonly used include SCF, TPO, Flt3L, IL-3, IL-6 and others, but no systematic studies have been carried out to analyze which of these factors are required, when they act, and what their effects on HSC specification are. Notably, the most successful approaches have used an in-vivo environment, which additionally provides multiple cell types for signaling molecules and extracellular matrix for three-dimensional mechanical support to developing HSCs. Taken together, this indicates that development currently can be directed most efficiently towards hematopoiesis by using the least defined conditions. Generation of virtually any system *de novo* however requires understanding of the natural state of the system, underscoring the need for further investigation of the late steps in the developing hematopoietic system.

Second, the environmental cues that maintain HSCs in their niche *in vivo* are poorly understood, and thus far no in-vitro culture system is able to sustain HSC self-renewal and/or potential. Consequently, even if, at some stage during a differentiation process, *bona fide* HSCs could be generated, they may exit the state of full differentiation and self-renewal potential between the moment of generation and the time point of assessment, leading to a false-negative outcome.

Most recently, large screening studies of chemical compound libraries and structure-function optimization have led to major progress in the formulation of a maintenance and expansion medium for HSC culture (Fares et al. 2014) (and unpublished data from Sauvageau laboratory). These novel advances have again fueled the hope that culture of HSCs *in vitro* without – or with minimal – loss of reconstitution ability is theoretically possible. The development of such a culture condition would mark a major milestone in the field of hematopoiesis.

It should be noted, however, that compared to PSCs, telomerase activity and DNA repair are considerably less active in HSCs, suggesting an inherent limit of self-renewal capacity (reviewed in (Naka and Hirao 2011)). Even *in vivo*, with a niche fully capable of optimally supporting HSC maintenance, serial transplantation of HSCs may not be unlimited (Kamminga et al. 2005; Sieburg et al. 2011), though conflicting evidence has been presented (Iscove and Nawa 1997). It should therefore be considered that indefinite expansion in culture may prove impossible, and that a maintenance medium, keeping HSCs, once isolated, in a quiescent, multipotent state would be almost equally beneficial in a research setting.

Acknowledgment We thank Dr. S. Tamir Rashid for careful reading of the manuscript and comments.

References

Amabile G, Welner RS, Nombela-Arrieta C et al (2013) In vivo generation of transplantable human hematopoietic cells from induced pluripotent stem cells. Blood 121:1255–1264. doi:10.1182/blood-2012-06-434407

Ambler CA, Nowicki JL, Burke AC, Bautch VL (2001) Assembly of trunk and limb blood vessels involves extensive migration and vasculogenesis of somite-derived angioblasts. Dev Biol 234:352–364. doi:10.1006/dbio.2001.0267

Anjos-Afonso F, Currie E, Palmer HG et al (2013) CD34(-) cells at the apex of the human hematopoietic stem cell hierarchy have distinctive cellular and molecular signatures. Cell Stem Cell 13:161–174. doi:10.1016/j.stem.2013.05.025

Batta K, Florkowska M, Kouskoff V, Lacaud G (2014) Direct reprogramming of murine fibroblasts to hematopoietic progenitor cells. Cell Rep 9:1871–1884. doi:10.1016/j.celrep.2014.11.002

Bertrand JY, Giroux S, Golub R et al (2005) Characterization of purified intraembryonic hematopoietic stem cells as a tool to define their site of origin. Proc Natl Acad Sci U S A 102:134–139. doi:10.1073/pnas.0402270102

Bertrand JY, Chi NC, Santoso B et al (2010) Haematopoietic stem cells derive directly from aortic endothelium during development. Nature 464:108–111. doi:10.1038/nature08738

Blazsek I, Chagraoui J, Péault B (2000) Ontogenic emergence of the hematon, a morphogenetic stromal unit that supports multipotential hematopoietic progenitors in mouse bone marrow. Blood 96:3763–3771

Boisset J-C, van Cappellen W, Andrieu-Soler C et al (2010) In vivo imaging of haematopoietic cells emerging from the mouse aortic endothelium. Nature 464:116–120. doi:10.1038/nature08764

Bonzanni N, Garg A, Feenstra KA et al (2013) Hard-wired heterogeneity in blood stem cells revealed using a dynamic regulatory network model. Bioinformatics 29:i80–i88. doi:10.1093/bioinformatics/btt243

Burt RK, Verda L, Kim D-A et al (2004) Embryonic stem cells as an alternate marrow donor source: engraftment without graft-versus-host disease. J Exp Med 199:895–904. doi:10.1084/jem.20031916

Buta C, David R, Dressel R et al (2013) Reconsidering pluripotency tests: do we still need teratoma assays? Stem Cell Res 11:552–562. doi:10.1016/j.scr.2013.03.001

Cerdan C, Rouleau A, Bhatia M (2004) VEGF-A165 augments erythropoietic development from human embryonic stem cells. Blood 103:2504–2512. doi:10.1182/blood-2003-07-2563

Chadwick K, Wang L, Li L et al (2003) Cytokines and BMP-4 promote hematopoietic differentiation of human embryonic stem cells. Blood 102:906–915. doi:10.1182/blood-2003-03-0832

Chambers SM, Fasano CA, Papapetrou EP et al (2009) Highly efficient neural conversion of human ES and iPS cells by dual inhibition of SMAD signaling. Nat Biotechnol 27:275–280. doi:10.1038/nbt.1529

Chen J, Lansford R, Stewart V et al (1993) RAG-2-deficient blastocyst complementation: an assay of gene function in lymphocyte development. Proc Natl Acad Sci U S A 90:4528–4532

Chen MJ, Yokomizo T, Zeigler BM et al (2009) Runx1 is required for the endothelial to haematopoietic cell transition but not thereafter. Nature 457:887–891. doi:10.1038/nature07619

Chew J-L, Loh Y-H, Zhang W et al (2005) Reciprocal transcriptional regulation of Pou5f1 and Sox2 via the Oct4/Sox2 complex in embryonic stem cells. Mol Cell Biol 25:6031–6046. doi:10.1128/MCB.25.14.6031-6046.2005

Choi K, Kennedy M, Kazarov A et al (1998) A common precursor for hematopoietic and endothelial cells. Development 125:725–732

Christensen JL, Weissman IL (2001) Flk-2 is a marker in hematopoietic stem cell differentiation: a simple method to isolate long-term stem cells. Proc Natl Acad Sci U S A 98:14541–14546. doi:10.1073/pnas.261562798

Christensen JL, Wright DE, Wagers AJ, Weissman IL (2004) Circulation and chemotaxis of fetal hematopoietic stem cells. PLoS Biol 2:E75. doi:10.1371/journal.pbio.0020075

Conlon FL, Lyons KM, Takaesu N et al (1994) A primary requirement for nodal in the formation and maintenance of the primitive streak in the mouse. Development 120:1919–1928

Corbel C, Salaün J, Belo-Diabangouaya P, Dieterlen-Lièvre F (2007) Hematopoietic potential of the pre-fusion allantois. Dev Biol 301:478–488. doi:10.1016/j.ydbio.2006.08.069

Cumano A, Dieterlen-Lièvre F, Godin I (1996) Lymphoid potential, probed before circulation in mouse, is restricted to caudal intraembryonic splanchnopleura. Cell 86:907–916

D'Amour KA, Agulnick AD, Eliazer S et al (2005) Efficient differentiation of human embryonic stem cells to definitive endoderm. Nat Biotechnol 23:1534–1541. doi:10.1038/nbt1163

Davidson AJ, Ernst P, Wang Y et al (2003) cdx4 mutants fail to specify blood progenitors and can be rescued by multiple hox genes. Nature 425:300–306. doi:10.1038/nature01973

Davis RP, Ng ES, Costa M et al (2008) Targeting a GFP reporter gene to the MIXL1 locus of human embryonic stem cells identifies human primitive streak-like cells and enables isolation of primitive hematopoietic precursors. Blood 111:1876–1884. doi:10.1182/blood-2007-06-093609

de Bruijn MF, Speck NA, Peeters MC, Dzierzak E (2000) Definitive hematopoietic stem cells first develop within the major arterial regions of the mouse embryo. EMBO J 19:2465–2474. doi:10.1093/emboj/19.11.2465

Doetschman TC, Eistetter H, Katz M et al (1985) The in vitro development of blastocyst-derived embryonic stem cell lines: formation of visceral yolk sac, blood islands and myocardium. J Embryol Exp Morphol 87:27–45

Doulatov S, Vo LT, Chou SS et al (2013) Induction of multipotential hematopoietic progenitors from human pluripotent stem cells via respecification of lineage-restricted precursors. Cell Stem Cell 13:459–470. doi:10.1016/j.stem.2013.09.002

Durand C, Robin C, Bollerot K et al (2007) Embryonic stromal clones reveal developmental regulators of definitive hematopoietic stem cells. Proc Natl Acad Sci U S A 104:20838–20843. doi:10.1073/pnas.0706923105

Dush MK, Martin GR (1992) Analysis of mouse Evx genes: Evx-1 displays graded expression in the primitive streak. Dev Biol 151:273–287

Ema H, Nakauchi H (2000) Expansion of hematopoietic stem cells in the developing liver of a mouse embryo. Blood 95:2284–2288

Evans MJ, Kaufman MH (1981) Establishment in culture of pluripotential cells from mouse embryos. Nature 292:154–156

Fares I, Chagraoui J, Gareau Y et al (2014) Cord blood expansion. Pyrimidoindole derivatives are agonists of human hematopoietic stem cell self-renewal. Science 345:1509–1512. doi:10.1126/science.1256337

Ferkowicz MJ, Starr M, Xie X et al (2003) CD41 expression defines the onset of primitive and definitive hematopoiesis in the murine embryo. Development 130:4393–4403

Fitch SR, Kimber GM, Wilson NK et al (2012) Signaling from the sympathetic nervous system regulates hematopoietic stem cell emergence during embryogenesis. Cell Stem Cell 11:554–566. doi:10.1016/j.stem.2012.07.002

Forlani S, Lawson KA, Deschamps J (2003) Acquisition of Hox codes during gastrulation and axial elongation in the mouse embryo. Development 130:3807–3819

Funayama N, Sato Y, Matsumoto K et al (1999) Coelom formation: binary decision of the lateral plate mesoderm is controlled by the ectoderm. Development 126:4129–4138

Gadue P, Huber TL, Paddison PJ, Keller GM (2006) Wnt and TGF-beta signaling are required for the induction of an in vitro model of primitive streak formation using embryonic stem cells. Proc Natl Acad Sci U S A 103:16806–16811. doi:10.1073/pnas.0603916103

Gekas C, Dieterlen-Lièvre F, Orkin SH, Mikkola HKA (2005) The placenta is a niche for hematopoietic stem cells. Dev Cell 8:365–375. doi:10.1016/j.devcel.2004.12.016

Gertow K, Wolbank S, Rozell B et al (2004) Organized development from human embryonic stem cells after injection into immunodeficient mice. Stem Cells Dev 13:421–435. doi:10.1089/1547328041797499

Giorgetti A, Montserrat N, Aasen T et al (2009) Generation of induced pluripotent stem cells from human cord blood using OCT4 and SOX2. Cell Stem Cell 5:353–357. doi:10.1016/j.stem.2009.09.008

Goodell MA, Brose K, Paradis G et al (1996) Isolation and functional properties of murine hematopoietic stem cells that are replicating in vivo. J Exp Med 183:1797–1806

Graf T, Enver T (2009) Forcing cells to change lineages. Nature 462:587–594. doi:10.1038/nature08533

Guo G, Huss M, Tong GQ et al (2010) Resolution of cell fate decisions revealed by single-cell gene expression analysis from zygote to blastocyst. Dev Cell 18:675–685. doi:10.1016/j.devcel.2010.02.012

Gupta S, Zhu H, Zon LI, Evans T (2006) BMP signaling restricts hemato-vascular development from lateral mesoderm during somitogenesis. Development 133:2177–2187. doi:10.1242/dev.02386

Hanna J, Wernig M, Markoulaki S et al (2007) Treatment of sickle cell anemia mouse model with iPS cells generated from autologous skin. Science 318:1920–1923. doi:10.1126/science.1152092

Hart AH, Hartley L, Sourris K et al (2002) Mixl1 is required for axial mesendoderm morphogenesis and patterning in the murine embryo. Development 129:3597–3608

Hole N, Graham GJ, Menzel U, Ansell JD (1996) A limited temporal window for the derivation of multilineage repopulating hematopoietic progenitors during embryonal stem cell differentiation in vitro. Blood 88:1266–1276

Huyhn A, Dommergues M, Izac B et al (1995) Characterization of hematopoietic progenitors from human yolk sacs and embryos. Blood 86:4474–4485

Ieda M, Fu J-D, Delgado-Olguin P et al (2010) Direct reprogramming of fibroblasts into functional cardiomyocytes by defined factors. Cell 142:375–386. doi:10.1016/j.cell.2010.07.002

Iscove NN, Nawa K (1997) Hematopoietic stem cells expand during serial transplantation in vivo without apparent exhaustion. Curr Biol 7:805–808

Ivanovs A, Rybtsov S, Anderson RA et al (2014) Identification of the niche and phenotype of the first human hematopoietic stem cells. Stem Cell Rep 2:449–456. doi:10.1016/j.stemcr.2014.02.004

Jokubaitis VJ, Sinka L, Driessen R et al (2008) Angiotensin-converting enzyme (CD143) marks hematopoietic stem cells in human embryonic, fetal, and adult hematopoietic tissues. Blood 111:4055–4063. doi:10.1182/blood-2007-05-091710

Kamminga LM, van Os R, Ausema A et al (2005) Impaired hematopoietic stem cell functioning after serial transplantation and during normal aging. Stem Cells 23:82–92. doi:10.1634/stemcells.2004-0066

Kaufman DS, Hanson ET, Lewis RL et al (2001) Hematopoietic colony-forming cells derived from human embryonic stem cells. Proc Natl Acad Sci U S A 98:10716–10721. doi:10.1073/pnas.191362598

Keller G, Kennedy M, Papayannopoulou T, Wiles MV (1993) Hematopoietic commitment during embryonic stem cell differentiation in culture. Mol Cell Biol 13:473–486

Kennedy M, Firpo M, Choi K et al (1997) A common precursor for primitive erythropoiesis and definitive haematopoiesis. Nature 386:488–493. doi:10.1038/386488a0

Kent DG, Copley MR, Benz C et al (2009) Prospective isolation and molecular characterization of hematopoietic stem cells with durable self-renewal potential. Blood 113:6342–6350. doi:10.1182/blood-2008-12-192054

Kiel MJ, Yilmaz OH, Iwashita T et al (2005) SLAM family receptors distinguish hematopoietic stem and progenitor cells and reveal endothelial niches for stem cells. Cell 121:1109–1121. doi:10.1016/j.cell.2005.05.026

Kikuchi K, Kondo M (2006) Developmental switch of mouse hematopoietic stem cells from fetal to adult type occurs in bone marrow after birth. Proc Natl Acad Sci U S A 103:17852–17857. doi:10.1073/pnas.0603368103

Kim DK, Fujiki Y, Fukushima T et al (1999) Comparison of hematopoietic activities of human bone marrow and umbilical cord blood CD34 positive and negative cells. Stem Cells 17:286–294. doi:10.1002/stem.170286

Kim JB, Greber B, Araúzo-Bravo MJ et al (2009) Direct reprogramming of human neural stem cells by OCT4. Nature 461:649–653. doi:10.1038/nature08436

Kinder SJ, Tsang TE, Quinlan GA et al (1999) The orderly allocation of mesodermal cells to the extraembryonic structures and the anteroposterior axis during gastrulation of the mouse embryo. Development 126:4691–4701

Kinder SJ, Tsang TE, Wakamiya M et al (2001) The organizer of the mouse gastrula is composed of a dynamic population of progenitor cells for the axial mesoderm. Development 128:3623–3634

Kispert A, Herrmann BG (1994) Immunohistochemical analysis of the Brachyury protein in wild-type and mutant mouse embryos. Dev Biol 161:179–193. doi:10.1006/dbio.1994.1019

Kissa K, Herbomel P (2010) Blood stem cells emerge from aortic endothelium by a novel type of cell transition. Nature 464:112–115. doi:10.1038/nature08761

Kobayashi T, Yamaguchi T, Hamanaka S et al (2010) Generation of rat pancreas in mouse by interspecific blastocyst injection of pluripotent stem cells. Cell 142:787–799. doi:10.1016/j.cell.2010.07.039

Kumano K, Chiba S, Kunisato A et al (2003) Notch1 but not Notch2 is essential for generating hematopoietic stem cells from endothelial cells. Immunity 18:699–711

Kumaravelu P, Hook L, Morrison AM et al (2002) Quantitative developmental anatomy of definitive haematopoietic stem cells/long-term repopulating units (HSC/RUs): role of the aorta-gonad-mesonephros (AGM) region and the yolk sac in colonisation of the mouse embryonic liver. Development 129:4891–4899

Kyba M, Perlingeiro RCR, Daley GQ (2002) HoxB4 confers definitive lymphoid-myeloid engraftment potential on embryonic stem cell and yolk sac hematopoietic progenitors. Cell 109:29–37

Lancrin C, Sroczynska P, Stephenson C et al (2009) The haemangioblast generates haematopoietic cells through a haemogenic endothelium stage. Nature 457:892–895. doi:10.1038/nature07679

Larochelle A, Vormoor J, Hanenberg H et al (1996) Identification of primitive human hematopoietic cells capable of repopulating NOD/SCID mouse bone marrow: implications for gene therapy. Nat Med 2:1329–1337

Lawson KA, Dunn NR, Roelen BA et al (1999) Bmp4 is required for the generation of primordial germ cells in the mouse embryo. Genes Dev 13:424–436

Le Douarin NM, Dieterlen-Lièvre F, Oliver PD (1984) Ontogeny of primary lymphoid organs and lymphoid stem cells. Am J Anat 170:261–299. doi:10.1002/aja.1001700305

Li W, Johnson SA, Shelley WC et al (2003) Primary endothelial cells isolated from the yolk sac and para-aortic splanchnopleura support the expansion of adult marrow stem cells in vitro. Blood 102:4345–4353. doi:10.1182/blood-2003-03-0729

Liu F, Kang I, Park C et al (2012) ER71 specifies Flk-1+ hemangiogenic mesoderm by inhibiting cardiac mesoderm and Wnt signaling. Blood 119:3295–3305. doi:10.1182/blood-2012-01-403766

Lu CC, Robertson EJ (2004) Multiple roles for Nodal in the epiblast of the mouse embryo in the establishment of anterior-posterior patterning. Dev Biol 273:149–159. doi:10.1016/j.ydbio.2004.06.004

Lu LS, Wang SJ, Auerbach R (1996) In vitro and in vivo differentiation into B cells, T cells, and myeloid cells of primitive yolk sac hematopoietic precursor cells expanded > 100-fold by coculture with a clonal yolk sac endothelial cell line. Proc Natl Acad Sci U S A 93:14782–14787

Lu S-J, Li F, Vida L, Honig GR (2004) CD34 + CD38- hematopoietic precursors derived from human embryonic stem cells exhibit an embryonic gene expression pattern. Blood 103:4134–4141. doi:10.1182/blood-2003-10-3575

Luckett WP (1978) Origin and differentiation of the yolk sac and extraembryonic mesoderm in presomite human and rhesus monkey embryos. Am J Anat 152:59–97. doi:10.1002/aja.1001520106

Maekawa M, Yamaguchi K, Nakamura T et al (2011) Direct reprogramming of somatic cells is promoted by maternal transcription factor Glis1. Nature 474:225–229. doi:10.1038/nature10106

Mahlapuu M, Ormestad M, Enerbäck S, Carlsson P (2001) The forkhead transcription factor Foxf1 is required for differentiation of extra-embryonic and lateral plate mesoderm. Development 128:155–166

Marshall CJ, Kinnon C, Thrasher AJ (2000) Polarized expression of bone morphogenetic protein-4 in the human aorta-gonad-mesonephros region. Blood 96:1591–1593

Marshall CJ, Sinclair JC, Thrasher AJ, Kinnon C (2007) Bone morphogenetic protein 4 modulates c-Kit expression and differentiation potential in murine embryonic aorta-gonad-mesonephros haematopoiesis in vitro. Br J Haematol 139:321–330. doi:10.1111/j.1365-2141.2007.06795.x

Matsumoto K, Isagawa T, Nishimura T et al (2009) Stepwise development of hematopoietic stem cells from embryonic stem cells. PLoS One 4, e4820. doi:10.1371/journal.pone.0004820

Matsuoka S, Tsuji K, Hisakawa H et al (2001) Generation of definitive hematopoietic stem cells from murine early yolk sac and paraaortic splanchnopleures by aorta-gonad-mesonephros region-derived stromal cells. Blood 98:6–12

Medvinsky AL, Samoylina NL, Müller AM, Dzierzak EA (1993) An early pre-liver intraembryonic source of CFU-S in the developing mouse. Nature 364:64–67. doi:10.1038/364064a0

Meier-Stiegen F, Schwanbeck R, Bernoth K et al (2010) Activated Notch1 target genes during embryonic cell differentiation depend on the cellular context and include lineage determinants and inhibitors. PLoS One 5, e11481. doi:10.1371/journal.pone.0011481

Mikkola HKA, Fujiwara Y, Schlaeger TM et al (2003) Expression of CD41 marks the initiation of definitive hematopoiesis in the mouse embryo. Blood 101:508–516. doi:10.1182/blood-2002-06-1699

Mitjavila MT, Filippi MD, Cohen-Solal K et al (1998) The Mpl-ligand is involved in the growth-promoting activity of the murine stromal cell line MS-5 on ES cell-derived hematopoiesis. Exp Hematol 26:124–134

Mitrani E, Ziv T, Thomsen G et al (1990) Activin can induce the formation of axial structures and is expressed in the hypoblast of the chick. Cell 63:495–501

Miyagi T, Takeno M, Nagafuchi H et al (2002) Flk1+ cells derived from mouse embryonic stem cells reconstitute hematopoiesis in vivo in SCID mice. Exp Hematol 30:1444–1453

Montserrat N, Nivet E, Sancho-Martinez I et al (2013) Reprogramming of human fibroblasts to pluripotency with lineage specifiers. Cell Stem Cell 13:341–350. doi:10.1016/j.stem.2013.06.019

Morita Y, Ema H, Nakauchi H (2010) Heterogeneity and hierarchy within the most primitive hematopoietic stem cell compartment. J Exp Med 207:1173–1182. doi:10.1084/jem.20091318

Morrison SJ, Wandycz AM, Akashi K et al (1996) The aging of hematopoietic stem cells. Nat Med 2:1011–1016

Müller AM, Medvinsky A, Strouboulis J et al (1994) Development of hematopoietic stem cell activity in the mouse embryo. Immunity 1:291–301

Mummery CL, Zhang J, Ng ES et al (2012) Differentiation of human embryonic stem cells and induced pluripotent stem cells to cardiomyocytes: a methods overview. Circ Res 111:344–358. doi:10.1161/CIRCRESAHA.110.227512

Najm FJ, Chenoweth JG, Anderson PD et al (2011) Isolation of epiblast stem cells from preimplantation mouse embryos. Cell Stem Cell 8:318–325. doi:10.1016/j.stem.2011.01.016

Naka K, Hirao A (2011) Maintenance of genomic integrity in hematopoietic stem cells. Int J Hematol 93:434–439. doi:10.1007/s12185-011-0793-z

Nakagawa M, Ichikawa M, Kumano K et al (2006) AML1/Runx1 rescues Notch1-null mutation-induced deficiency of para-aortic splanchnopleural hematopoiesis. Blood 108:3329–3334. doi:10.1182/blood-2006-04-019570

Nichols J, Smith A (2009) Naive and primed pluripotent states. Cell Stem Cell 4:487–492. doi:10.1016/j.stem.2009.05.015

Nishikawa SI, Nishikawa S, Hirashima M et al (1998) Progressive lineage analysis by cell sorting and culture identifies FLK1 + VE-cadherin + cells at a diverging point of endothelial and hemopoietic lineages. Development 125:1747–1757

North TE, de Bruijn MFTR, Stacy T et al (2002) Runx1 expression marks long-term repopulating hematopoietic stem cells in the midgestation mouse embryo. Immunity 16:661–672

Nostro MC, Cheng X, Keller GM, Gadue P (2008) Wnt, activin, and BMP signaling regulate distinct stages in the developmental pathway from embryonic stem cells to blood. Cell Stem Cell 2:60–71. doi:10.1016/j.stem.2007.10.011

Notta F, Doulatov S, Laurenti E et al (2011) Isolation of single human hematopoietic stem cells capable of long-term multilineage engraftment. Science 333:218–221. doi:10.1126/science.1201219

Nottingham WT, Jarratt A, Burgess M et al (2007) Runx1-mediated hematopoietic stem-cell emergence is controlled by a Gata/Ets/SCL-regulated enhancer. Blood 110:4188–4197. doi:10.1182/blood-2007-07-100883

Oberlin E, Tavian M, Blazsek I, Péault B (2002) Blood-forming potential of vascular endothelium in the human embryo. Development 129:4147–4157

Osawa M, Hanada K, Hamada H et al (1996) Long-term lymphohematopoietic reconstitution by a single CD34-low/negative hematopoietic stem cell. Science 273:242–245. doi:10.1126/science.273.5272.242

Ottersbach K, Dzierzak E (2005) The murine placenta contains hematopoietic stem cells within the vascular labyrinth region. Dev Cell 8:377–387. doi:10.1016/j.devcel.2005.02.001

Palis J, Robertson S, Kennedy M et al (1999) Development of erythroid and myeloid progenitors in the yolk sac and embryo proper of the mouse. Development 126:5073–5084

Palis J, Chan RJ, Koniski A et al (2001) Spatial and temporal emergence of high proliferative potential hematopoietic precursors during murine embryogenesis. Proc Natl Acad Sci U S A 98:4528–4533. doi:10.1073/pnas.071002398

Park C, Afrikanova I, Chung YS et al (2004) A hierarchical order of factors in the generation of FLK1- and SCL-expressing hematopoietic and endothelial progenitors from embryonic stem cells. Development 131:2749–2762. doi:10.1242/dev.01130

Pereira C-F, Chang B, Qiu J et al (2013) Induction of a hemogenic program in mouse fibroblasts. Cell Stem Cell 13:205–218. doi:10.1016/j.stem.2013.05.024

Pick M, Azzola L, Mossman A et al (2007) Differentiation of human embryonic stem cells in serum-free medium reveals distinct roles for bone morphogenetic protein 4, vascular endothelial growth factor, stem cell factor, and fibroblast growth factor 2 in hematopoiesis. Stem Cells 25:2206–2214. doi:10.1634/stemcells.2006-0713

Pulecio J, Nivet E, Sancho-Martinez I et al (2014) Conversion of human fibroblasts into monocyte-like progenitor cells. Stem Cells 32:2923–2938. doi:10.1002/stem.1800

Rao CV, Wolf DM, Arkin AP (2002) Control, exploitation and tolerance of intracellular noise. Nature 420:231–237. doi:10.1038/nature01258

Rhodes KE, Gekas C, Wang Y et al (2008) The emergence of hematopoietic stem cells is initiated in the placental vasculature in the absence of circulation. Cell Stem Cell 2:252–263. doi:10.1016/j.stem.2008.01.001

Riddell J, Gazit R, Garrison BS et al (2014) Reprogramming committed murine blood cells to induced hematopoietic stem cells with defined factors. Cell 157:549–564. doi:10.1016/j.cell.2014.04.006

Robert-Moreno A, Espinosa L, la Pompa de JL, Bigas A (2005) RBPjkappa-dependent Notch function regulates Gata2 and is essential for the formation of intra-embryonic hematopoietic cells. Development 132:1117–1126. doi:10.1242/dev.01660

Robin C, Durand C (2010) The roles of BMP and IL-3 signaling pathways in the control of hematopoietic stem cells in the mouse embryo. Int J Dev Biol 54:1189–1200. doi:10.1387/ijdb.093040cr

Robin C, Bollerot K, Mendes S et al (2009) Human placenta is a potent hematopoietic niche containing hematopoietic stem and progenitor cells throughout development. Cell Stem Cell 5:385–395. doi:10.1016/j.stem.2009.08.020

Sandler VM, Lis R, Liu Y et al (2014) Reprogramming human endothelial cells to haematopoietic cells requires vascular induction. Nature 511:312–318. doi:10.1038/nature13547

Sasaki H, Hogan BL (1993) Differential expression of multiple fork head related genes during gastrulation and axial pattern formation in the mouse embryo. Development 118:47–59

Sato Y (2013) Dorsal aorta formation: separate origins, lateral-to-medial migration, and remodeling. Dev Growth Differ 55:113–129. doi:10.1111/dgd.12010

Schmitt TM, de Pooter RF, Gronski MA et al (2004) Induction of T cell development and establishment of T cell competence from embryonic stem cells differentiated in vitro. Nat Immunol 5:410–417. doi:10.1038/ni1055

Sekiya S, Suzuki A (2011) Direct conversion of mouse fibroblasts to hepatocyte-like cells by defined factors. Nature 475:390–393. doi:10.1038/nature10263

Sharp JG, Kessinger A, Lynch JC et al (2000) Blood stem cell transplantation: factors influencing cellular immunological reconstitution. J Hematother Stem Cell Res 9:971–981. doi:10.1089/152581600750062435

Shi Y, Kirwan P, Smith J et al (2012) Human cerebral cortex development from pluripotent stem cells to functional excitatory synapses. Nat Neurosci 15:477–86– S1. doi:10.1038/nn.3041

Shultz LD, Lyons BL, Burzenski LM et al (2005) Human lymphoid and myeloid cell development in NOD/LtSz-scid IL2R gamma null mice engrafted with mobilized human hemopoietic stem cells. J Immunol 174:6477–6489

Sieburg HB, Rezner BD, Muller-Sieburg CE (2011) Predicting clonal self-renewal and extinction of hematopoietic stem cells. Proc Natl Acad Sci U S A 108:4370–4375. doi:10.1073/pnas.1011414108

Soares ML, Haraguchi S, Torres-Padilla M-E et al (2005) Functional studies of signaling pathways in peri-implantation development of the mouse embryo by RNAi. BMC Dev Biol 5:28. doi:10.1186/1471-213X-5-28

Solnica-Krezel L, Sepich DS (2012) Gastrulation: making and shaping germ layers. Annu Rev Cell Dev Biol 28:687–717. doi:10.1146/annurev-cellbio-092910-154043

Sumi T, Tsuneyoshi N, Nakatsuji N, Suemori H (2008) Defining early lineage specification of human embryonic stem cells by the orchestrated balance of canonical Wnt/beta-catenin, Activin/Nodal and BMP signaling. Development 135:2969–2979. doi:10.1242/dev.021121

Suzuki N, Yamazaki S, Yamaguchi T et al (2013) Generation of engraftable hematopoietic stem cells from induced pluripotent stem cells by way of teratoma formation. Mol Ther 21:1424–1431. doi:10.1038/mt.2013.71

Szabo E, Rampalli S, Risueño RM et al (2010) Direct conversion of human fibroblasts to multilineage blood progenitors. Nature 468:521–526. doi:10.1038/nature09591

Takahashi K, Yamanaka S (2006) Induction of pluripotent stem cells from mouse embryonic and adult fibroblast cultures by defined factors. Cell 126:663–676. doi:10.1016/j.cell.2006.07.024

Takahashi K, Tanabe K, Ohnuki M et al (2007) Induction of pluripotent stem cells from adult human fibroblasts by defined factors. Cell 131:861–872. doi:10.1016/j.cell.2007.11.019

Takashima Y, Guo G, Loos R et al (2014) Resetting transcription factor control circuitry toward ground-state pluripotency in human. Cell 158:1254–1269. doi:10.1016/j.cell.2014.08.029

Takeuchi M, Sekiguchi T, Hara T et al (2002) Cultivation of aorta-gonad-mesonephros-derived hematopoietic stem cells in the fetal liver microenvironment amplifies long-term repopulating activity and enhances engraftment to the bone marrow. Blood 99:1190–1196

Tavian M, Hallais MF, Péault B (1999) Emergence of intraembryonic hematopoietic precursors in the pre-liver human embryo. Development 126:793–803

Tavian M, Robin C, Coulombel L, Péault B (2001) The human embryo, but not its yolk sac, generates lympho-myeloid stem cells: mapping multipotent hematopoietic cell fate in intraembryonic mesoderm. Immunity 15:487–495

Tesar PJ, Chenoweth JG, Brook FA et al (2007) New cell lines from mouse epiblast share defining features with human embryonic stem cells. Nature 448:196–199. doi:10.1038/nature05972

Theunissen TW, Powell BE, Wang H et al (2014) Systematic identification of culture conditions for induction and maintenance of naive human pluripotency. Cell Stem Cell 15:471–487. doi:10.1016/j.stem.2014.07.002

Thomson JA, Itskovitz-Eldor J, Shapiro SS et al (1998) Embryonic stem cell lines derived from human blastocysts. Science 282:1145–1147

Tonegawa A, Takahashi Y (1998) Somitogenesis controlled by Noggin. Dev Biol 202:172–182. doi:10.1006/dbio.1998.8895

Tonegawa A, Funayama N, Ueno N, Takahashi Y (1997) Mesodermal subdivision along the mediolateral axis in chicken controlled by different concentrations of BMP-4. Development 124:1975–1984

Utikal J, Maherali N, Kulalert W, Hochedlinger K (2009) Sox2 is dispensable for the reprogramming of melanocytes and melanoma cells into induced pluripotent stem cells. J Cell Sci 122:3502–3510. doi:10.1242/jcs.054783

Vierbuchen T, Ostermeier A, Pang ZP et al (2010) Direct conversion of fibroblasts to functional neurons by defined factors. Nature 463:1035–1041. doi:10.1038/nature08797

Vijayaragavan K, Szabo E, Bossé M et al (2009) Noncanonical Wnt signaling orchestrates early developmental events toward hematopoietic cell fate from human embryonic stem cells. Cell Stem Cell 4:248–262. doi:10.1016/j.stem.2008.12.011

Vodyanik MA, Bork JA, Thomson JA, Slukvin II (2005) Human embryonic stem cell-derived CD34+ cells: efficient production in the coculture with OP9 stromal cells and analysis of lymphohematopoietic potential. Blood 105:617–626. doi:10.1182/blood-2004-04-1649

Wang L, Li L, Shojaei F et al (2004) Endothelial and hematopoietic cell fate of human embryonic stem cells originates from primitive endothelium with hemangioblastic properties. Immunity 21:31–41. doi:10.1016/j.immuni.2004.06.006

Wang L, Menendez P, Shojaei F et al (2005) Generation of hematopoietic repopulating cells from human embryonic stem cells independent of ectopic HOXB4 expression. J Exp Med 201:1603–1614. doi:10.1084/jem.20041888

Weintraub H, Tapscott SJ, Davis RL et al (1989) Activation of muscle-specific genes in pigment, nerve, fat, liver, and fibroblast cell lines by forced expression of MyoD. Proc Natl Acad Sci U S A 86:5434–5438

Wiles MV, Keller G (1991) Multiple hematopoietic lineages develop from embryonic stem (ES) cells in culture. Development 111:259–267

Wilson NK, Foster SD, Wang X et al (2010) Combinatorial transcriptional control in blood stem/progenitor cells: genome-wide analysis of ten major transcriptional regulators. Cell Stem Cell 7:532–544. doi:10.1016/j.stem.2010.07.016

Wineman J, Moore K, Lemischka I, Müller-Sieburg C (1996) Functional heterogeneity of the hematopoietic microenvironment: rare stromal elements maintain long-term repopulating stem cells. Blood 87:4082–4090

Woll PS, Morris JK, Painschab MS et al (2008) Wnt signaling promotes hematoendothelial cell development from human embryonic stem cells. Blood 111:122–131. doi:10.1182/blood-2007-04-084186

Woods N-B, Parker AS, Moraghebi R et al (2011) Brief report: efficient generation of hematopoietic precursors and progenitors from human pluripotent stem cell lines. Stem Cells 29:1158–1164. doi:10.1002/stem.657

Xie H, Ye M, Feng R, Graf T (2004) Stepwise reprogramming of B cells into macrophages. Cell 117:663–676

Yamamoto R, Morita Y, Ooehara J et al (2013) Clonal analysis unveils self-renewing lineage-restricted progenitors generated directly from hematopoietic stem cells. Cell 154:1112–1126. doi:10.1016/j.cell.2013.08.007

Yoder MC, Hiatt K, Dutt P et al (1997) Characterization of definitive lymphohematopoietic stem cells in the day 9 murine yolk sac. Immunity 7:335–344

Yoon M-J, Koo B-K, Song R et al (2008) Mind bomb-1 is essential for intraembryonic hematopoiesis in the aortic endothelium and the subaortic patches. Mol Cell Biol 28:4794–4804. doi:10.1128/MCB.00436-08

Yu J, Vodyanik MA, Smuga-Otto K et al (2007) Induced pluripotent stem cell lines derived from human somatic cells. Science 318:1917–1920. doi:10.1126/science.1151526

Zambidis ET, Péault B, Park TS et al (2005) Hematopoietic differentiation of human embryonic stem cells progresses through sequential hematoendothelial, primitive, and definitive stages resembling human yolk sac development. Blood 106:860–870. doi:10.1182/blood-2004-11-4522

Zape JP, Zovein AC (2011) Hemogenic endothelium: origins, regulation, and implications for vascular biology. Semin Cell Dev Biol 22:1036–1047. doi:10.1016/j.semcdb.2011.10.003

Zeigler BM, Sugiyama D, Chen M et al (2006) The allantois and chorion, when isolated before circulation or chorio-allantoic fusion, have hematopoietic potential. Development 133:4183–4192. doi:10.1242/dev.02596

Zhou X, Sasaki H, Lowe L et al (1993) Nodal is a novel TGF-beta-like gene expressed in the mouse node during gastrulation. Nature 361:543–547. doi:10.1038/361543a0

Chapter 2
Generation and Regeneration of T Cells

Hiroshi Kawamoto, Tomokatsu Ikawa, Takuya Maeda, and Kyoko Masuda

Abstract T cells contain a variety of lineages, each of which is formed by passing through a number of strictly regulated steps. In this article we aim to clarify essential factors and critical developmental steps during T cell development. First, we will describe the process of T cell development that occurs *in vivo*. Then, by presenting recent approaches in synthetic biology, we will show that, in the murine case, a feeder-free culture system using a combination of cytokines and Notch ligand is sufficient to support T cell development starting from multipotent hematopoietic progenitors to the TCR-expressing $CD4^+CD8^+$ stage. Finally, in the human case, we will show that mature functional $CD8^+$ killer T cells can be produced *in vitro* starting as early as from the ES/iPS cell stage using feeder cells. These studies may help clarify "minimal requirements" for T cell development.

Abbreviations

HSC	hematopoietic stem cell
TCR	T cell receptor
MLP	myelo-lymphoid progenitor
LMPP	lymphoid-primed multipotent progenitor
DLL	delta-like ligand

H. Kawamoto (✉) • K. Masuda
Department of Immunology, Institute for Frontier Medical Science, Kyoto University, Kyoto 606-8507, Japan
e-mail: kawamoto@frontier.kyoto-u.ac.jp

T. Ikawa
Laboratory for Immune Regeneration, RIKEN Center for Integrative Medical Sciences, Yokohama 230-0045, Japan

T. Maeda
Department of Immunology, Institute for Frontier Medical Science, Kyoto University, Kyoto 606-8507, Japan

Department of Hematology and Oncology, Graduate School of Medicine, Kyoto University, Kyoto 606-8507, Japan

T. Watanabe, Y. Takahama (eds.), *Synthetic Immunology*,
DOI 10.1007/978-4-431-56027-2_2

TEC	thymic epithelial cell
iLS cells	induced leukocyte stem cell
iPSC	induced pluripotent stem cell
CTL	cytotoxic T lymphocyte
TIL	tumor-infiltrating lymphocyte

Introduction

During early development, multipotent hematopoietic stem cells (HSCs) are stepwisely restricted to eventually become unipotent T cell progenitors. T cell progenitors then rearrange their T cell receptor genes and are subsequently strictly selected in the thymus based on T cell receptor (TCR) quality before maturation. In the physiological situation, developing progenitors sequentially encounter different environments, whereby they receive appropriate developmental cues *in situ*. It seems almost impossible to faithfully recapitulate the complexity of events *in vitro*. However, not all environmental factors are necessarily indispensable, and there is a possibility that T cells can be differentiated in a more simplified fashion. This idea can be straightforwardly tested in *in vitro* culture systems, and by doing so, factors that are essential for T cell development should be revealed.

Here, we first explain the process of lineage restriction from HSCs to T cell progenitors, based mainly on our previous work. Then, we focus on the steps that occur during early intrathymic T cell development. Finally, we discuss to what extent it is possible to recapitulate T cell development *in vitro* for both mouse and human cells.

Developmental Pathway from HSCs to T Cell Progenitors

Proposal of Myeloid Based Model

Most hematology and immunology textbooks describe that the first branch point from HSCs produces two progenitors, namely one for myelo-erythroid cells and another one for lymphoid cells including T and B cells (Fig. 2.1a). This model is based on the concept that the blood cell family can be subdivided into two major lineages, a myelo-erythroid lineage and a lymphoid lineage. In 2001, we proposed an alternative model in which myeloid potential is retained at an early stage of branching toward erythroid, T and B cell lineages (Fig. 2.1b) (Katsura and Kawamoto 2001), which we later named myeloid-based model in 2006 (Kawamoto 2006).

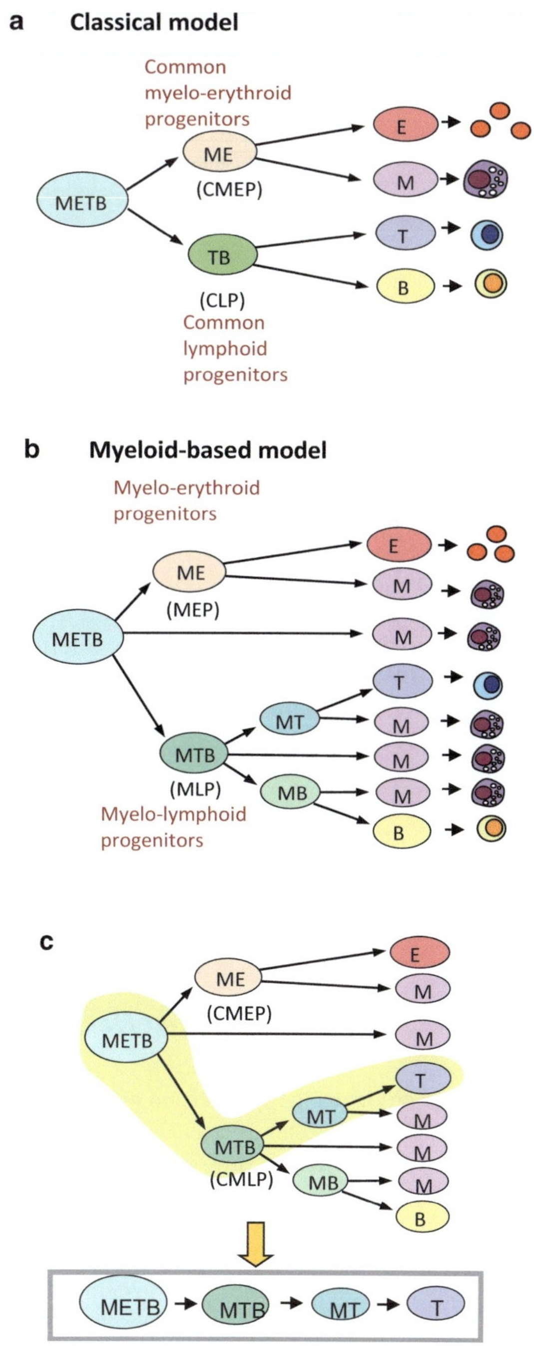

Fig. 2.1 Representative models of hematopoiesis (**a**) The classical model. This model proposes that the HSC first diverges into a CMEP and a CLP. It should be noted that CMEPs are sometimes referred to as common myeloid progenitors (CMPs). E, M, T, and B indicate the progenitor potential for erythroid, myeloid, T, and B cells, respectively. *CMEP* common myelo-erythorid progenitor, *CLP* common lymphocyte progenitor (**b**) Myeloid-based model. In this model, the first branch point generates MEPs and MLPs and the myeloid potential persists in the T and B cell branches even after these lineages have diverged. *MEP* myelo-erythorid progenitor; *MLP* myelo-lymphoid progenitor (**c**) Stepwise lineage restriction from multipotent progenitor toward T cell progenitor

Myeloid Based Model as a Map of Developmental Potentials

Cell differentiation models have two different aspects: they serve as a map of developmental potential and as a cell fate map. In other words, these two are expressed as "a map for lineage restriction" and "a map for physiological production routes". We argue that a map of developmental potential is first and foremost essential for the study of molecular mechanisms of lineage commitment.

In this article, we focus specifically on the route from HSCs to T cell progenitors (Fig. 2.1c). In the myeloid-based model, multipotent progenitors (M-E-T-B progenitors) are initially restricted to become myelo-lymphoid progenitors (MLPs) (or M-T-B progenitors). Subsequently, MLPs terminate B cell potential to form myeloid-T progenitors (M-T progenitors), and finally myeloid-T progenitors shut off myeloid potential to become T cell progenitors.

Myelo-Lymphoid Progenitor Stage (M-T-B Progenitor Status)

MLPs were first identified at a single cell level in fetal liver by our group in 2002 (Lu et al. 2002), and in bone marrow by Adolfsson et al. (2005). These cells are also called LMPPs (lymphoid-primed multipotent progenitors). The existence of this stage in murine and human hematopoiesis has been confirmed by many groups (Yoshida et al. 2006; Pronk et al. 2007; Doulatov et al. 2010). Further study showed that MLPs retain the potential to produce neutrophils, while myelo-erythroid progenitors are generating eosinophils and basophils (Gorgens et al. 2013). Although the contribution of MLPs to myeloid cell production *in vivo* is still a matter of debate, the presence of a commitment status having M-T-B potential seems indisputable.

Prethymic Pathway of T Cell Development

T cells are produced in thymus from progenitors that immigrate from hematopoietic organs, specifically fetal liver during the fetal period and bone marrow after birth. The nature of thymus-seeding progenitors has long been disputed. Regarding the earliest thymic immigrants, we have shown that they are biased toward the T cell lineage. In 1997, we developed a clonal assay system able to detect T, B and myeloid lineage potential of individual progenitors (Kawamoto et al. 1997). By using this assay, a progenitor that generated only T cells was determined as "p-T (progenitor-T)". Such p-Ts were subsequently detected in the aorta-gonad-mesonephros (AGM) region (Ohmura et al. 1999), and fetal liver (Kawamoto et al. 1997, 1998). However, p-Ts determined by a "retrospective" method do not necessarily represent T cell lineage restricted progenitors, since the tested progenitors do not always express their full potential in culture. Nevertheless, later we provided convincing evidence for the presence of T cell progenitors at the prethymic stage by showing that p-Ts can be highly enriched in the $Lin^{-}c\text{-}kit^{+}IL\text{-}7R^{+}$ population of fetal liver (Kawamoto et al. 2000) and fetal blood (Ikawa et al. 2004a). We and another group also showed that progenitors that are closely juxtaposed to the thymic epithelial primordium were T cell lineage restricted (Masuda et al. 2005a; Harman et al. 2005), validating that the earliest thymic immigrants are T cell lineage restricted. We further disclosed that these prethymic T cell progenitors specifically express paired immunoglobulin-like receptors (PIR) (Fig. 2.2) (Masuda et al. 2005b). The expression of PIR is immediately down-regulated upon entrance to the fetal thymus, substantiating the prethymic stage as a distinct stage. While we had referred to these prethymic progenitors as "T cell lineage restricted" progenitors, in our later studies we came to know that such prethymic progenitors retain significant myeloid potential and also residual B cell potential (Kawamoto 2006). We may thus call them "T-biased MLPs" at present.

Early studies have demonstrated that thymic migration takes place in two waves during fetal development (Jotereau et al. 1987; Dunon et al. 1999). Recently, it has been shown that the second wave-progenitors colonizing thymus just before birth retain robust B cell potential in addition to myeloid potential (Luc et al. 2012; Ramond et al. 2014). It is therefore probable that the second wave-progenitors are close to MLPs. On the other hand, characteristics of thymus-seeding progenitors in adult thymus still remain unclear.

Myeloid-T Progenitors

We then investigated whether myeloid or B cell potential is retained longer during the process of T cell development in adult thymus. This question was resolved by analyzing whether or not early thymic T cell progenitors retain myeloid potential,

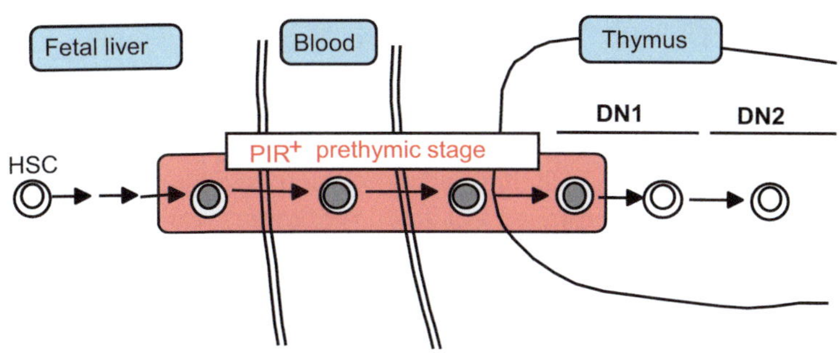

Fig. 2.2 Prethymic pathway of T cell development can be defined by the expression of PIR $Lin^{-}c\text{-}kit^{+}IL\text{-}7R^{+}PIR^{+}$ cells are present in fetal liver, fetal blood and fetal thymus at an early stage during the fetal period. These cells represent myeloid-T progenitors, while they retain residual B cell potential. *HSC* hematopoietic stem cell, *DN* double negative

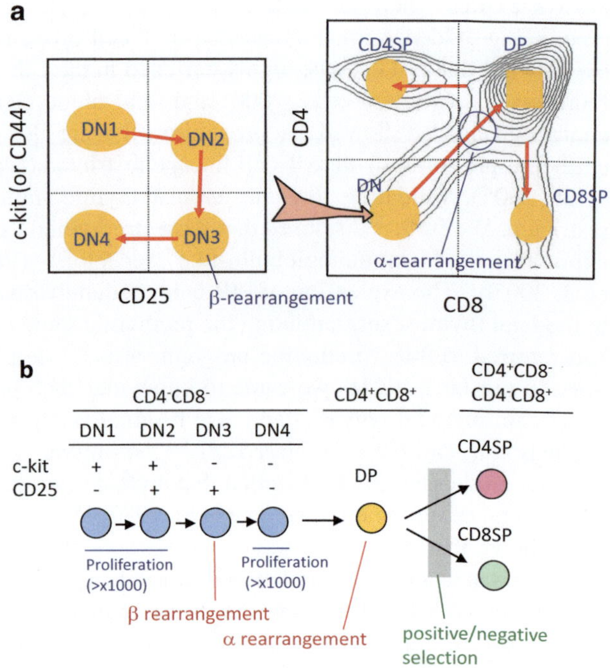

Fig. 2.3 Intrathymic T cell development
(**a**) The sequence of intrathymic developmental stages, which are defined by flow cytometric profiles, is shown
(**b**) Schematic illustration of intrathymic developmental stages

because earlier we and other groups had demonstrated that T cell progenitors in the earliest population in adult thymus have mostly lost B cell potential (Porritt et al. 2004; Balciunaite et al. 2005; Lu et al. 2005).

Figure 2.3a, b explain the definition of developmental stages of murine thymocytes. The early $CD4^-CD8^-$ (DN: double negative) stage of thymocytes can be further subdivided into DN1, DN2, DN3, and DN4 stages, based on c-kit vs CD25 expression profiles, with the DN1 stage representing the earliest stage. DN cells then differentiate into the $CD4^+CD8^+$ (DP: double positive) stage, where thymocytes come to express αβTCR. DP cells undergo positive and negative selection to become mature $CD4^+CD8^-$ or $CD4^-CD8^+$ (SP: single positive) cells.

By analyzing individual DN1 cells in a culture system which can support generation of both T and myeloid cells, we found that a substantial proportion of thymic T cell progenitors produced macrophages in addition to T cells (Wada et al. 2008) (Fig. 2.4a, b). A similar finding was reported by Bell and Bhandoola (2008). These results demonstrated that T cell progenitors retain myeloid potential after terminating B cell potential (Fig. 2.4c). Therefore, it became clear that a commitment status producing myeloid cells and T cells, but not B cells, exists on the developmental pathway from HSCs to T cell progenitors, strongly arguing against the classical model of hematopoiesis.

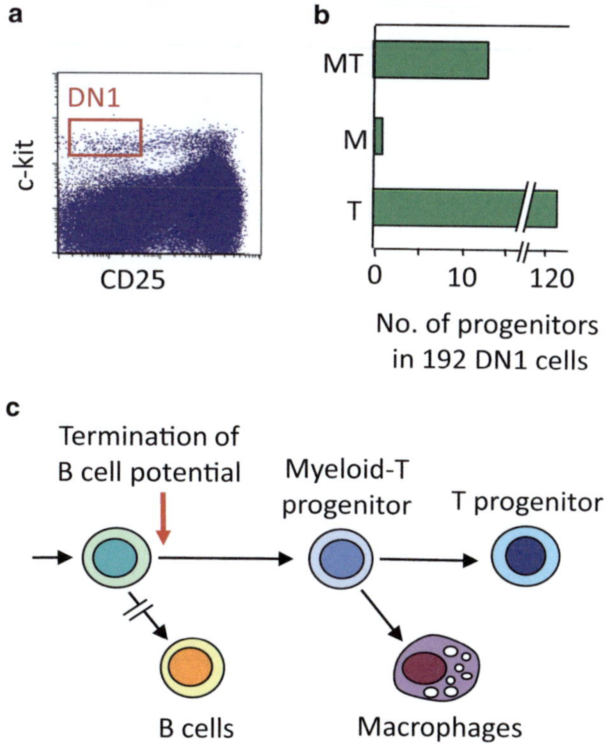

Fig. 2.4 T cell progenitors in adult thymus that have lost B cell potential retain macrophage potential (**a**) c-kit vs CD25 profile of the $CD3^-CD4^-CD8^-$ fraction of adult thymus cells from 8-week-old mice (**b**) A total of 192 individual DN1 cells were cultured with TSt-4/DLL1 that can support generation of T cells and myeloid cells. Seeded cells were retrospectively classified according to the cells generated in each clonal culture: progenitors that generated both macrophages and T (MT), only macrophages (M) and only T cells (T) (**c**) Early T cell progenitors in the adult thymus that have lost B cell potential still retain a significant ability to generate macrophages

Myeloid Potential of T Cell Progenitors is Terminated in the Midst of the DN2 Stage

We also examined at which step the myeloid potential of M-T progenitors is terminated to become fully T cell lineage restricted. We showed that in plckGFP Tg mice, where GFP expression is driven by the proximal *lck* promoter, DN2 cells can be subdivided into GFP^- and GFP^+ cells (Masuda et al. 2007) (Fig. 2.5a). We found that GFP^-DN2 cells retain the potential to produce non-T lineage cells, including macrophages, dendritic cells, and natural killer cells, while GFP^+DN2 cells failed to produce such non-T cells (Wada et al. 2008; Masuda et al. 2007). We recently designated these two stages DN2mt (myeloid-T) and DN2t (T-lineage determined) (Fig. 2.5b) and termed the step between these stages the DN2-determination step (Kawamoto et al. 2010). Rothenberg et al. also paid attention to this step and proposed to subdivide the DN2 stage into DN2a and DN2b stages based on c-kit expression levels (DN2a: c-$kit^{high}CD25^{high}$, DN2b: c-$kit^{mid}CD25^{high}$) (Rothenberg et al. 2008). We presumed that the DN2-determination step should be a critical developmental checkpoint for the production of the T cell lineage.

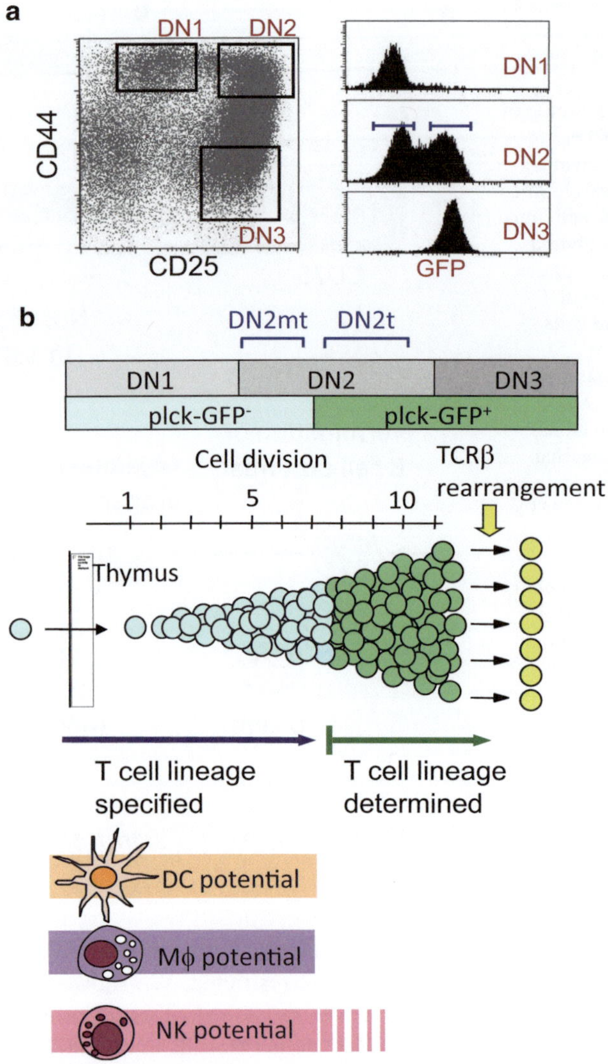

Fig. 2.5 Myeloid potential is terminated in the midst of the DN2 stage
(**a**) CD44 vs CD25 profile of the Lin^{-} fraction of thymocytes from 15 dpc (days post coitum) fetuses of plckGFP Tg mice is shown in the left panel. The right panels show GFP expression by cells in the gates denoted in the left panel
(**b**) Schematic illustration of the early differentiation/proliferation of thymocytes. A single early thymic progenitor undergoes more than 10 cell divisions during the DN1 and DN2 stages to generate more than 1000 DN3 cells. The shut-off of non-T lineage potential, including that for dendritic cells (DC), macrophages (Mϕ), and natural killer (NK) cells, occurs during the transition from the GFP^{-}DN2 stage to GFP^{+}DN2 stage, and subsequently the T cell lineage determined progenitors undergo several cell divisions before they enter the DN3 stage to initiate TCRβ gene rearrangement. We designate GFP^{-}DN2 and GFP^{+}DN2 stages as DN2mt (myeloid-T) and DN2t (T-lineage determined) stages, respectively

Identification of a Developmental Checkpoint for Production of the T Cell Lineage

A reliable way to substantiate that a presumed step is critical for the development of a certain cell lineage is to demonstrate that developmental arrest occurs at the stage prior to that step under defined conditions. If the arrested cells begin self-renewal, then the existence of a critical step becomes clear.

To visualize such critical step, we used a feeder-free culture system that had been modified from the one originally developed by Bernstein's group (Varnum-Finney et al. 2003), in which murine hematopoietic progenitors were cultured on immobilized Notch ligand DLL4 molecules in the presence of a cocktail of cytokines including IL-7 (Fig. 2.6a). We found that progenitors developing toward T cells were arrested at the DN2mt stage (Fig. 2.6b), and that the arrested cells entered a self-renewal cycle maintaining non-T lineage potentials (Ikawa et al. 2010). A similar arrest and self-renewal of progenitors was observed in thymocytes of mice deficient in the transcription factor Bcl11b (Fig. 2.6c), indicating that Bcl11b plays a critical role in the DN2-determination step. The importance of Bcl11b in the T cell lineage determination step was also demonstrated by other groups (Li et al. 2010a, b). Fig. 2.6d summarizes the T cell lineage determination step that occurs at the DN2mt-DN2t transition step, which is Bcl11b-dependent.

Previously, it has been reported that Gata3 is absolutely indispensable for the production of T cells, and a developmental block was thought to take place at the prethymic stage (Ting et al. 1996). However, to date, it remains unclear at which stage the developmental arrest occurs. Tcf-1, together with Lef-1 that exhibits functional redundancy with Tcf-1, has also been known as a critical regulator of T cell development (Okamura et al. 1998). Recent studies showed that Tcf-1 starts to play a role at the DN1 stage (Weber et al. 2011); however, it may not be critical for T cell lineage determination, since Tcf1$^{-/-}$ Lef-1$^{-/-}$ thymocytes were shown to develop into the DN4 stage (Yu et al. 2012).

Environmental Factors That Support T Cell Development

Factors Required for Thymic Colonization

For the earliest thymic immigrants, three chemokines, namely CCL21, CCL25, and CXCL12, are essential to colonizing the not yet vascularized thymic anlage (Liu et al. 2006; Calderon and Boehm 2011). Interestingly, CCL21 is expressed in the parathyroid anlage (Liu et al. 2006), which is not separated from the thymic anlage at this stage (12 days post coitum). This finding is in line with previous observations showing that colonizing progenitors are attracted close to the fetal thymic anlage of nude mouse, in which parathyroid anlage is normally formed (Itoi et al. 2001). These three molecules may be also important for thymic immigration via blood

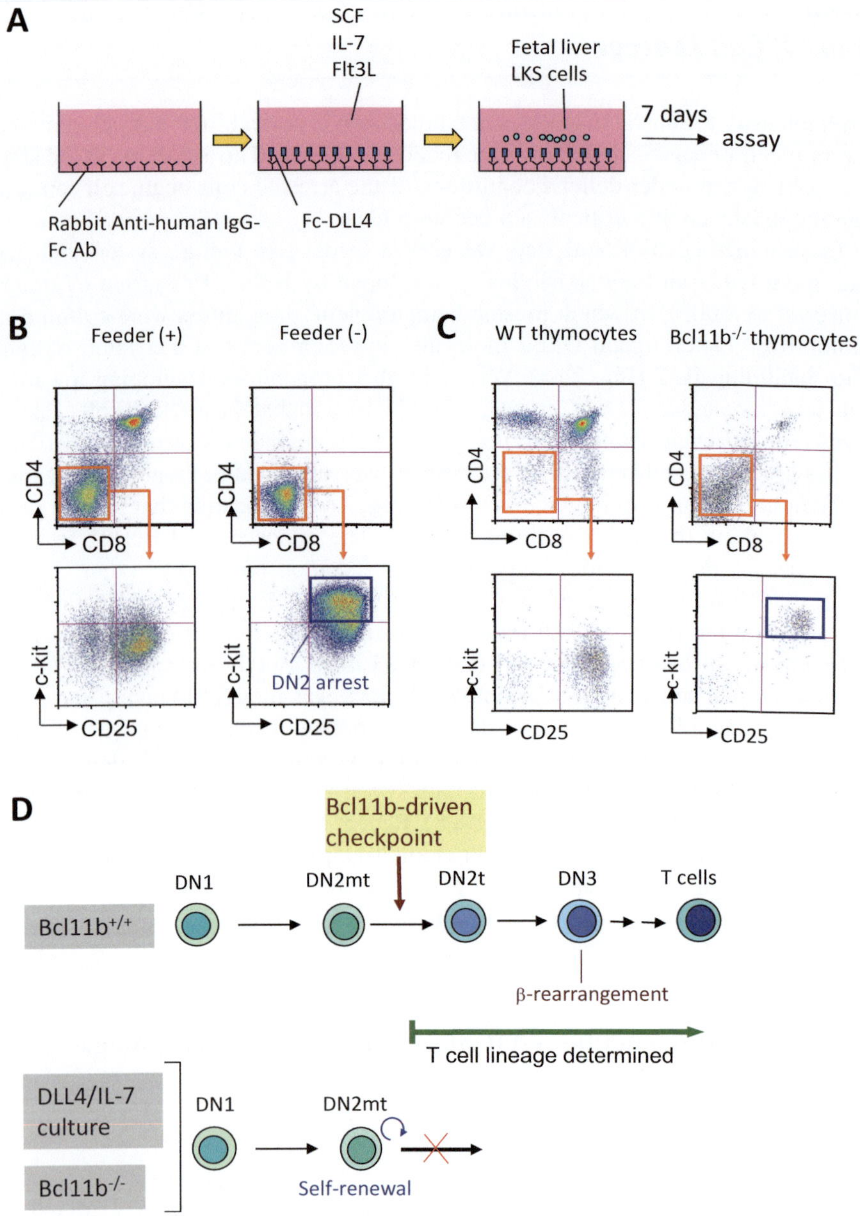

Fig. 2.6 The step for the termination of myeloid potential serves as a critical developmental checkpoint in T cell development

(**a**) Schematic illustration of the feeder-free culture system. A 96-well plate was coated with rabbit anti-human IgG-Fc antibody. Then, Fc-DLL4 fusion proteins were added and immobilized to the dish. $Lin^{-}ckit^{+}Sca1^{+}$ (LKS) cells (200 cells) from murine fetal liver were cultured with immobilized Fc-DLL4 in the presence of SCF, IL-7 and Flt3L

(**b**) In the feeder-free condition, cells are developmentally arrested at the DN2 stage. Left lanes show the cells generated using feeder cells. LKS cells from murine fetal liver mice were cultured

vessels, since thymocyte numbers are drastically reduced in fetuses deficient in CCR7, CCR9 and CXCR4 (receptors for CCL21, CCL25, and CXCL12, respectively) even several days after vascularization (Calderon and Boehm 2011). It was also demonstrated that P-selectin and its ligand PSGL-1 are important for thymic colonization in adult thymus (Rossi et al. 2005).

Notch Signal is Essential for T Cell Development

Early T cell development has been shown to be dependent on IL-7 and stem cell factor (Peschon et al. 1994; von Freeden-Jeffry et al. 1995; Rodewald et al. 1997). However, expression of these factors is not specific for thymus. Thus, the exclusive production of T cells in thymus has long been enigmatic, and researchers have been seeking for key environmental cues that induce exclusive T cell production in thymus.

In 1999, Notch was revealed to be such a key molecule. Conditional knock out of Notch1 resulted in a complete loss of T cells in thymus (Radtke et al. 1999), while enforced expression of a constitutively active form of Notch1 in hematopoietic progenitors brought about production of T lineage cells in bone marrow (Pui et al. 1999). Later, it was shown that a major ligand for Notch1 in thymus is Delta like 4 (DLL4) (Hozumi et al. 2008; Koch et al. 2008).

Expression of Unique Peptides in Cortical Thymic Epithelial Cells (cTECs) is Required for Positive Selection

DP cells have to pass positive selection in order to differentiate into SP cells. Positive selection is thought to select T cells that can recognize the complex of MHC and exogenous antigen, whereas negative selection is thought to deplete harmful autoreactive T cells. As mechanisms for these selections, it has been believed that T cells receiving a moderate signal by the MHC-self-peptide complex are positively selected, while those receiving a strong signal by the MHC-self-

Fig. 2.6 (continued) with TSt-4/DLL4 cells for 7 days. Right lanes show cells generated on day 7 in the feeder-free condition indicated in (**a**)
(**c**) Bcl11b is essential for T cell lineage determination. Fetal liver cells from Bcl11b$^{+/+}$ or Bcl11b$^{-/-}$ mice (Ly5.2) were transferred into lethally irradiated mice (Ly5.1). Flow cytometric profiles of reconstituted thymocytes of recipient mice 8 weeks after transfer are shown. In the lower panels, c-kit vs CD25 profiles of cells gated on the CD3^{-}CD4^{-}CD8^{-} fraction are shown
(**d**) A scheme of T cell lineage determination. Non-T lineage potential of T cell progenitors is terminated during the DN2 stage, subdividing this stage into DN2mt and DN2t stages (see also Fig. 2.5). Bcl11b is required for the DN2mt uncommitted cells to pass through this checkpoint

peptide complex are deleted. Such difference in TCR signal strength has been thought to result from a differential reactivity of TCR that is formed by random gene recombination.

However, recent findings seem to require revision of the above concept. "Cathepsine L" is a lysosomal cysteine protease specifically expressed in cortical thymic epithelial cells and thus is thought to play a role in the presentation of cTEC-specific peptide on MHC-class II. In 2002, it was shown that CD4SP cells are reduced in cathepsine L deficient mice (Honey et al. 2002). Later, another type of cTEC-specific protease, TSSP (thymus-specific serine protease), has also been demonstrated to be involved in processing peptide for positive selection of CD4 T cells (Viret et al. 2011).

A similar finding was reported for CD8 T cells. In 2007, a cTEC-specific proteasome component, β5t, was found, and its deletion was shown to result in marked reduction of CD8 T cells (Murata et al. 2007). A very recent study has shown that OTI-TCR expressing T cells generated in β5t-deficient thymus were functionally inferior to those generated in wild-type thymus, indicating that the positive selection process not only allows survival of T cells but also determines functional maturation of T cells (Takada et al. 2015).

These findings indicate that cTECs have to express unique peptides for positive selection of thymocytes. It is thus difficult to explain above findings in the light of a classical selection model that presumes that the difference in TCR-MHC signal strength is exclusively attributable to TCR diversity. Regarding the role of such unique peptides proper to cTECs, two models have been proposed. In model (i), cTEC-specific peptides preferentially generate moderate signals to TCR. In model (ii), cTECs merely induce positive selection for reactive T cells, while medullary TECs (mTECs) and medullary dendritic cells induce negative selection for reactive cells. In model (ii), a certain number of cells survive because cTECs express a set of peptides different from that of mTECs. Using the latter model, the phenotype of cathepsine L- or β5t-deficient mice can be explained by the assumption that when both cTECs and mTECs express the same repertoire of peptides, only few T cells can survive. In contrast, recently reported findings that peptides processed by the β5t-containing proteasome tend to exhibit low affinity for TCR tend to support the former model (Sasaki et al. 2015). Notwithstanding the possibility of either model, the basic mechanisms regulating positive selection still remain to be elucidated.

Approaches Based on the Concept of Synthetic Biology

In Vivo *Recapitulation of the Phylogeny of Thymus*

In nude mouse, thymus fails to form. Foxn1 has been shown to be the molecule responsible for this athymic phenotype (Nehls et al. 1994). However, even in nude mice, thymic anlage is formed in the correct location (Itoi et al. 2001). Recently,

T. Boehm's group genetically modified Foxn1-deficient mice so that TECs come to express a specified number of molecules (Calderon and Boehm 2012). Among them, two major chemokines, CCL25 and CXCL12 that were shown to work as thymic attractants (Liu et al. 2006; Calderon and Boehm 2011), were selected in addition to SCF and DLL4. IL-7 was not used in this experiment, since Foxn1-deficient TECs express IL-7. None of these four molecules solely reconstituted thymopoiesis, although CCL25 or CXCL12 attracted hematopoietic progenitors and induced myeloid cell generation in thymus. Combination of chemokines and SCF lead to B cell generation in thymus. Addition of DLL4 then recapitulated generation of T lineage cells. Despite the fact that cellularity is much lower in such reconstituted thymus DP cells were formed. Although these findings may provide some insight into the events that occurred during thymus evolution, it seems that the findings themselves were mostly within expectations based on the results obtained in *in vitro* culture systems.

In Vitro *Recapitulation of T Cell Development*

Fetal Thymic Organ Culture

Thymic epithelial cells support T cell development in thymus. Then, is it possible to culture thymic epithelial cells *in vitro* and use them as feeder cells for T cells? It is indeed possible to culture thymic epithelial cells in a monolayer fashion. However, such monolayer-cultured thymic epithelial cells immediately lose their ability to support T cell development (our unpublished findings).

To circumvent the loss of TEC function in monolayer culture, fetal thymic lobes have been used in order to experimentally induce T cell development *in vitro*. In this method, fetal thymic lobes are first incubated with deoxyguanosine (dGuo), a treatment that depletes developing thymocytes but keeps stromal components alive (Jenkinson et al. 1982). Then, dGuo is removed by washing and hematopoietic progenitors are cultured together with the lobe. This method is very powerful because T cell development is fully supported (Ueno et al. 2005).

Co-Culture with Stromal Cells

Mesenchymal stromal cells derived from bone marrow are able to support the development of B cells and myeloid cells (Nishikawa et al. 1988; Nakano et al. 1994), but they are not capable of supporting T cell development. As mentioned earlier, Notch signal was found to be essential for T cell development (Radtke et al. 1999; Pui et al. 1999). Based on this finding, a method to culture T cells *in vitro* was developed in 2002 (Schmitt and Zuniga-Pflucker 2002). In this method, hematopoietic progenitors are cultured with OP-9 cells that were genetically engineered to over-express DLL1. Using this setup, it is possible to induce DP cells from hematopoietic progenitors. This co-culture system has been extensively used to study T cell development.

Besides its usefulness for the study of murine T cells, this co-culture system can equally well support human T cell development. It was shown that DP cells can be induced from cord blood $CD34^+$ cells (La Motte-Mohs et al. 2005). Some groups have proposed to use this culture system for the expansion of T cell progenitors that are to be transferred to humans in the clinical setting (Awong et al. 2009; Meek et al. 2010). For example, patients who undergo bone marrow transplantation often face the problem that reconstitution of T cells from donor-derived stem/progenitor cells tends to be delayed, and that such delayed T cell production may lead to an immune-compromised status. In this case, it is expected that *in vitro* cultured T cell progenitors may efficiently colonize the thymus and thus accelerate reconstitution of the recipient's immune function.

Developmental Arrest-Restart Controlling Culture System

Here, we show several cases where we used a stromal co-culture system in order to elucidate mechanisms of cell fate decisions during hematopoiesis. We thought that, if we can recapitulate differentiation of progenitors *in vitro* in a "synchronous" manner, time course sampling of these cells would give us a lot of information regarding gene expression and epigenetic profiles.

To this aim, we first tried to develop a culture system in which we can manipulate arrest and restart of progenitor development. Recently, E. Rotherberg's group performed a comprehensive gene expression analysis and epigenetic profiling along with DN1-DN2a-DN2b-DN3-DP stage analysis and revealed counteraction of Gata3 vs PU.1 during lineage specification in terms of T vs myeloid lineage program (Zhang et al. 2012). We then decided to focus in more detail on the DN1 stage, where MLPs colonizing the thymus immediately terminate B cell potential and initiate the program toward the DN2 stage upon encountering thymic environments. We then paid attention to our previous finding that hematopoietic progenitors deficient in E2A, one of the E-protein family transcription factors essential for B cell development, give rise to self-renewing MLPs when they are cultured under B cell inducing conditions (Ikawa et al. 2004b). Progenitors deficient in PAX5 and EBF1 have been shown to form self-renewing MLPs as well (Nutt et al. 1999; Pongubala et al. 2008).

Based on these findings, we came to think that the simple blockade of differentiation should be sufficient to generate stem cells, i.e., to endow MLPs with self-renewing capability. In line with this idea, we have recently shown that overexpression of Id3, which dominant-negatively inhibits E protein activity, in murine multipotent hematopoietic progenitors leads to the generation of stem cells with T, B and myeloid potentials (Fig. 2.7) (Ikawa et al. 2015). We call these progenitors induced leukocyte stem (iLS) cells. We also showed that this blockade can be done in a more inducible way by using ER-Id3 transgene. There, iLS cells are produced by adding Tamoxifen to the culture medium. Then, by removing it

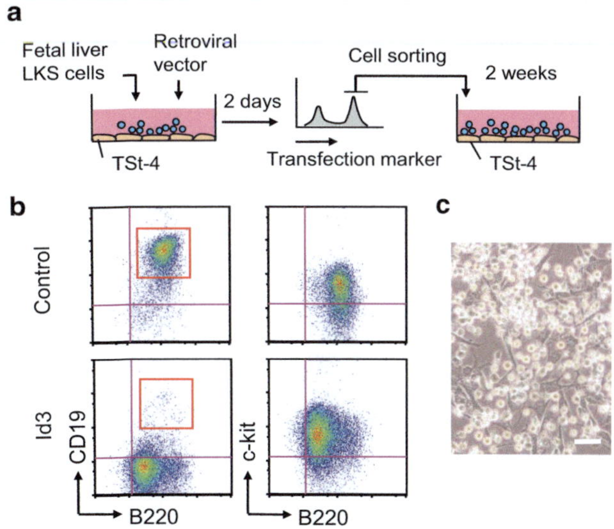

Fig. 2.7 Production of induced leukocyte stem cells
(**a**) Schematic representation of Id3-induced hematopoietic progenitor cell generation. $Lin^{-}ckit^{+}$ $Sca1^{+}$ (LKS) cells from murine fetal liver were transfected with retrovirus containing the Id3 gene, and transfected cells were further cultured with TSt-4 cells
(**b**) Flow cytometric analysis of control (empty vector) and Id3-overexpressing fetal liver progenitor cells. These Id3-overexpressing cells are developmentally arrested and enter a self-renewal cycle, retaining T, B and myeloid potentials. We term them induced leukocyte stem (iLS) cells
(**c**) Photomicrograph of iLS cells. The white bar indicates 10 μm

again, cells restart differentiation toward B cells. In addition, this method can be used for the arrest-restart system controlling the step from MLPs toward B cell lineage.

These findings may also give us new fundamental insights about stem cells. Self-renewal potential and multipotency are hallmarks of stem cells, and it is generally thought that the acquisition of such "stemness" requires rejuvenation of somatic cells through reprogramming of their genetic and epigenetic status. In contrast to this, our results propose that a simple blockade of cell differentiation is sufficient to induce and maintain stem cells.

Recapitulation of DN1 Stage Development Using the Arrest-Restart Controlling Culture System

We then tried to use self-renewing multipotent progenitors as a source of starting cells for synchronized differentiation culture toward T cells. For this purpose, E2A-deficient progenitors are not suitable, because E2A is also required for T cell development. We chose EBF1-deficient progenitors rather than PAX5-deficient progenitors, as EBF1-deficient progenitors are thought to be more immature.

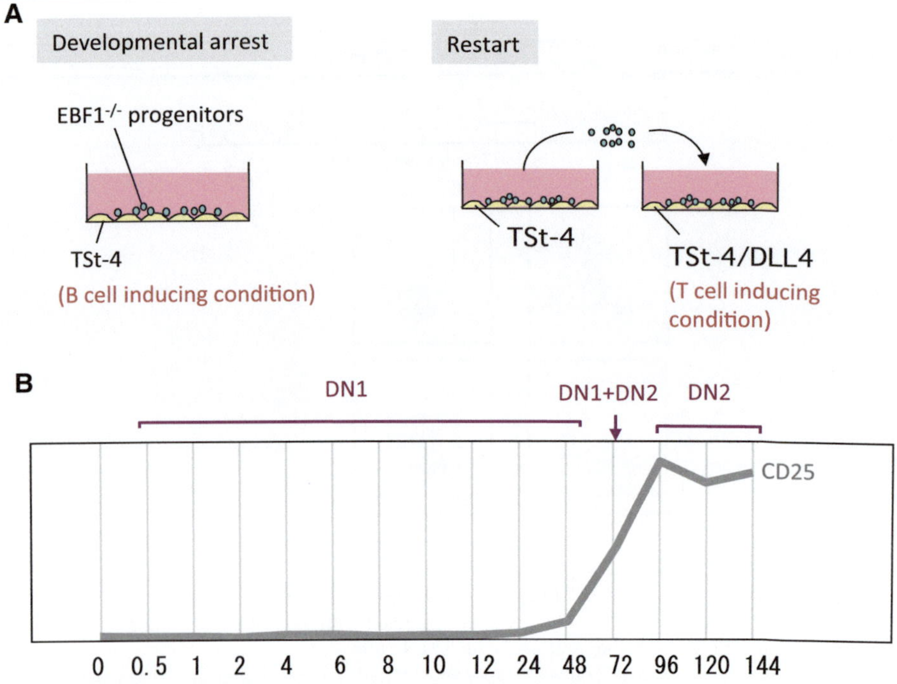

Fig. 2.8 Recapitulation of DN1 stage development in an *in vitro* culture system
(**a**) When EBF-1-deficient hematopoietic progenitors are maintained in co-culture with TSt-4 cells, cells are developmentally arrested and enter a self-renewal cycle, retaining the potential to produce T, B, and myeloid lineages. To induce differentiation toward the T cell lineage, these EBF-1-deficient multipotent progenitors were transferred to a co-culture with TSt-4/DLL1 cells (**b**) Transferred cells come to express CD25 on day 3 (72 h), and fully become DN2 cells on day 4. Time course samples were provided to the RIKEN FANTOM 5 project, and data of CAGE analysis for these samples are available online (http://fantom.gsc.riken.jp/zenbu/)

Using EBF1-deficient progenitors, we developed a culture system that recapitulated DN1–DN2 stages. EBF1-deficient progenitors can be maintained as self-renewing MLPs in a co-culture with TSt-4 cells optimal for B cell development (Fig. 2.8a, left panel). We then transferred EBF1-deficient progenitors to a co-culture with TSt-4/DLL1 cells, that is, TSt-4 cells that overexpress the Notch ligand DLL1 (Fig. 2.8a, right panel). On day 3 after transfer, cells came to express CD25, and on day 4 the entire cells became c-kit$^+$CD25$^+$ DN2 cells. Thus, transferred cells undergo differentiation in a very synchronous manner, and the first 2 days of culture are thought to recapitulate the DN1 stage where thymus-seeding progenitors terminate B cell potential upon encountering Notch ligand. In collaboration with the RIKEN FANTOM 5 project we prepared and analyzed a total of 15 time point RNA samples. FANTOM 5 performed CAGE analysis, which sequences the first 20 nucleic acids of cDNA in a quantitative manner by next generation sequencing, making it possible to quantify gene transcription levels and determine whole genome transcriptional starting sites. In line with flow cytometric analysis, CD25 started to be expressed on day 3 (Fig. 2.8b). The data have been

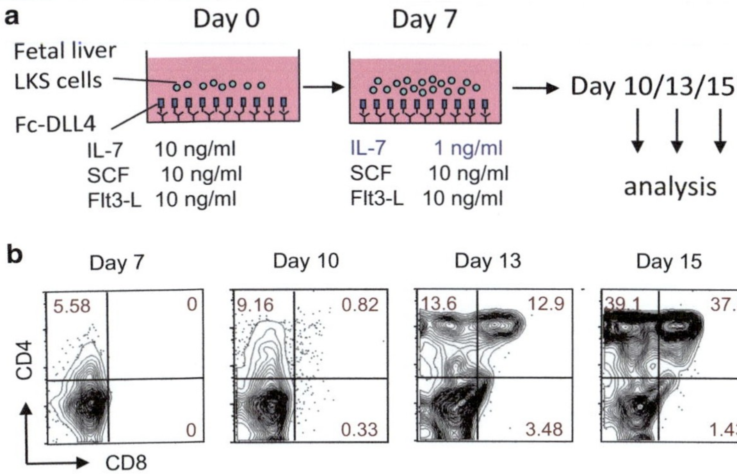

Fig. 2.9 Reduction of IL-7 concentration induces DP cell generation
(**a**) LKS cells sorted from 13 dpc fetal liver were cultured with immobilized DLL4 in the presence of SCF, IL-7 and Flt3L (10 ng/ml). The concentration of IL-7 was reduced to 1 ng/ml, and the cells were further cultured
(**b**) Kinetics of T cell generation in the immobilized Fc-DLL4 cultures

published (Arner et al. 2015) and are now available on the RIKEN FANTOM 5 "zenbu" site (http://fantom.gsc.riken.jp/zenbu/).

Feeder-Free Culture System

Stromal cells produce a myriad of undefined factors. Therefore, it is preferable to develop a culture system that does not use feeder cells. In 2003, it was reported that immobilized DLL1 and cytokines including IL-7 can support T cell development to the CD25$^+$ stage (Varnum-Finney et al. 2003), but it seemed difficult to induce further development.

As we have mentioned earlier, in our hand cells were also arrested at the DN2 stage in feeder-free culture initially (Fig. 2.6a, b). However, when we systematically optimized the feeder-free culture system, we found that arrested cells initiate differentiation when the concentration of IL-7 is reduced on day 7 of culture (10 ng/ml to 1 ng/ml) (Ikawa et al. 2010) (Fig. 2.9a). Notably, cells in these cultures developed up to the αβTCR-expressing DP stage (Fig. 2.9b). These results demonstrate that αβTCR$^+$ cells can be generated from multipotent hematopoietic progenitors in a "feeder-free" culture system, merely using cytokines and Notch ligand. Therefore, any cellular interactions are dispensable for T cell development, at least until the DP stage. Of interest was the finding that once cells reached the DN3 stage and started TCRβ chain rearrangement, DP cells were generated without changing culture conditions. This finding indicates that so called TCRβ-selection, which is thought to serve as a critical checkpoint for preTCR formation, does not require additional environmental factors in this feeder-free culture system.

Toward Clinical Application

Currently Ongoing Efficient Strategies in Cancer Immunotherapy

Recently, our group is trying to apply our *in vitro* culture methods to cancer immunotherapy. Over the past several years, cancer immunotherapy has remarkably progressed. One of the leading achievements is the strategy to block inhibitory signals in the immune system. CTLA-4 and PD-1 are expressed on activated T cells as inhibitory receptors. By blocking these receptors with specific monoclonal antibodies (mAbs), it is possible to augment the immune response. Indeed, anti-PD1 mAb as well as anti-CTLA4 mAb alone or in combination have been clinically shown to be effective for some types of cancers, including melanoma, lung cancer, and kidney cancer, even at the advanced stage (Hodi et al. 2010; Topalian et al. 2012; Wolchok et al. 2013). The approach using immune checkpoint blockade is thus thought to be an epoch-making breakthrough in cancer therapy. Nevertheless, several problems remain to be solved. One issue is that autoimmune reactions take place at a certain frequency since this strategy activate the immune system in a non-antigen-specific manner. Another issue is that, in most cases, the effect is limited to ~20 % of patients.

Another effective method in cancer immunotherapy is to directly use CTLs. S. A. Rosenberg's group has been taking the strategy in which tumor-infiltrating lymphocytes (TILs) are collected from the patient, activated *in vitro*, and then returned to the patient (Rosenberg et al. 1988). Recently, the group is taking a more aggressive strategy. In this strategy, the patient firstly undergoes chemotherapy and lethal irradiation to deplete pre-existing lymphocytes, followed by autologous HSC transplantation, and then TILs are transferred back to the patient (Rosenberg 2011). An obvious effect was seen in 70 % of metastatic melanoma patients, 40 % of whom survived more than 5 years.

TCR gene transfer is another effective approach, where peripheral T cells of a patient are transduced with a defined TCR gene using a retrovirus or a lentivirus vector system. For example, in clinical trail of MART-1-specific TCR gene against melanoma, 10–30 % of patients showed tumor regression (Morgan et al. 2006; Johnson et al. 2009). In another trial, T cells transduced with NY-ESO1-specific TCR have been shown to be effective in 60 % of synovial cell sarcoma patients and 40 % of melanoma patients (Robbins et al. 2011).

Cloning of Antigen-Specific T Cells by Using iPSC Technology

Although the checkpoint-blocking mAbs are effective in various types of cancer, autoimmune reactions inevitably occur at a certain frequency, as mentioned earlier. Direct use of CTLs has been also shown to be effective, but invasive pretreatment

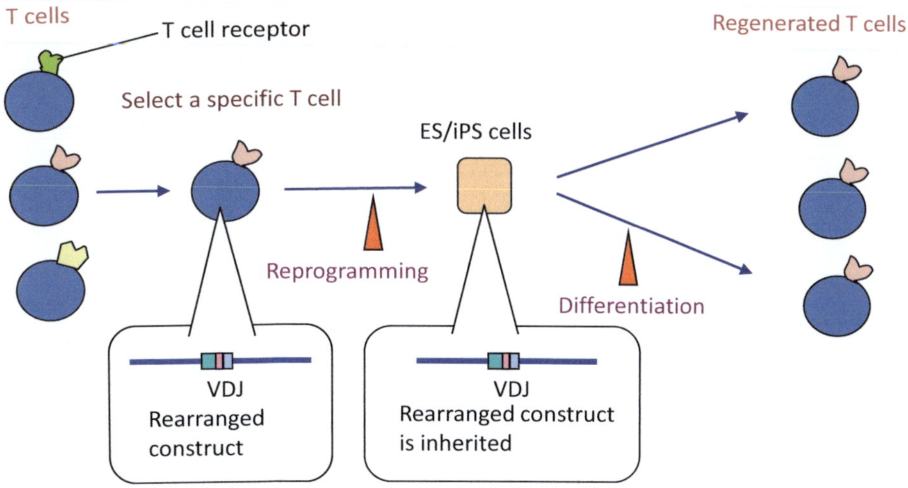

Fig. 2.10 Reprogramming of T cells with certain antigen specificities into iPSCs serves as a method of cloning
iPSCs produced by reprogramming of a T cell having a defined antigen specificity inherit rearranged genomic constructs of TCR genes of the original T cell. All T cells regenerated from such iPSCs are expected to express the same TCR as the original T cell

or gene therapy is required for the method to work well. The question may arise why one does not simply expand antigen-specific T cells *in vitro* by using tumor antigens and antigen presenting cells. Actually, antigen-specific CTLs can be expanded *in vitro*, but it has been extremely difficult to get a sufficient number of cells, since *in vitro* expanded CTLs tend to become easily exhausted after a certain period of culture.

To overcome this problem, we came to think of the idea to utilize induced pluripotent stem cell (iPSC) technology for the cloning and *in vitro* expansion of CTLs. Essentially iPSCs are established from mature antigen-specific T cells. As T cell receptors are formed from rearranged TCR genes, iPSCs derived from a T cell (T-iPSCs) should inherit those rearranged genomic structures (Fig. 2.10). All T cells regenerated from T-iPSCs are thus expected to express the same TCR on their surface. Since iPSCs can be expanded almost unlimitedly, it is possible to obtain as many fresh CTLs as needed.

Production of iPSCs from Tumor Antigen-Specific CTLs

Pursuing this idea, we have recently succeeded in establishing iPSCs from melanoma antigen MART1-specific CTLs originally derived from a melanoma patient,

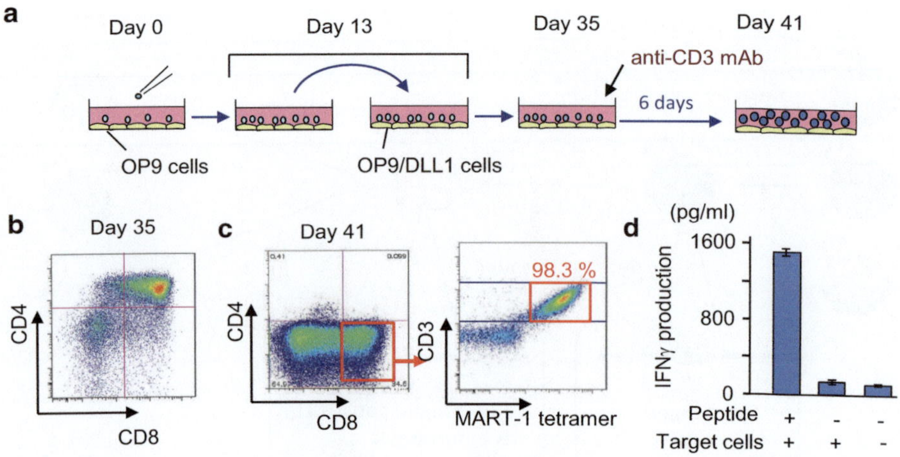

Fig. 2.11 Regeneration of MART-1-specific CTLs from MART1-T-iPSCs
(**a**) MART1-T-iPSCs were sequentially cultured with two types of feeder cells, OP9 and OP9/DLL1. On day 35 of cultivation, anti-CD3 mAb was added to induce the generation of mature T cells
(**b**) CD4/CD8 double positive cells were generated on day 35 of cultivation
(**c**) A large number of CD8 single positive cells were generated 6 days after the stimulation with anti-CD3 mAb. Virtually all of them expressed a TCR specific for the MART-1 antigen
(**d**) Production of IFNγ by regenerated $CD8^+$ T cells upon antigen-specific stimulation. IFNγ secretion by $CD8^+$ T cells was measured by ELISA using the supernatant after co-culturing 1×10^5 regenerated $CD8^+$ T cells for 24 h with 1×10^4 HLA-A*02:01-positive EBV-lymphoblastoid cells (CIRA0201) pulsed or not pulsed with MART-1 peptide (EAAGIGILTV). Mean ± SE of triplicates is shown

and in regenerating MART1-specific CTLs from these iPSCs (Vizcardo et al. 2013).

As a cell source we used JKF6 cells (Yang et al. 2011), which are specific for the melanoma antigen MART-1. JKF6 cells are long-term cultured TILs originally derived from a melanoma patient. Some previous studies reported that iPSCs can be produced from human peripheral T cells by using Yamanaka factors (Seki et al. 2010; Loh et al. 2010). However, in our case we also used SV40 (Park et al. 2008) to increase reprogramming efficiency. For transduction, we used the Sendai virus system (Fusaki et al. 2009). We thus succeeded in making iPSCs from MART1-specific CTLs (MART1-T-iPSCs).

Regeneration of Tumor Antigen-Specific CTLs from iPSCs

We then induced differentiation of T cells from MART1-T-iPSCs (Fig. 2.11a) by modification of a previously published method (Timmermans et al. 2009). After 35 days of cultivation, $CD4^+CD8^+$ DP cells were generated (Fig. 2.11b). However, these *in vitro* culture systems have been unable to support the efficient generation of SP cells beyond the DP stage.

To induce further differentiation, we simply added anti-CD3 mAb to the culture on day 35. The proportion of $CD8^+$ T cells clearly increased during a period of 6 days following TCR stimulation (Fig. 2.11c, left panel), and cells expanded by 300-fold (data not shown). Importantly, the resulting $CD8^+$ T cells were almost exclusively specific for the MART-1 antigen (Fig. 2.11c, right panel). Thus, in this culture system positive selection of $CD8^+$ T cells can be recapitulated *in vitro*.

To examine whether these $CD8^+$ T cells can be activated in an antigen-specific manner, regenerated $CD8^+$ T cells were co-cultured with target cells (human EBV-lymphoblastoid cell line) with or without MART-1 peptide. $CD8^+$ T cells produced a substantial amount of IFNγ only in the presence of MART-1 peptide (Fig. 2.11d). Collectively, this T-iPSC approach appears to be efficient in regenerating functional antigen-specific CTLs.

We recently found that DP cells expressing antigen-specific TCR can be induced to $CD8^+$ T cells by co-culturing with antigen presenting cells pulsed with agonist peptide (our unpublished observation). Such apparent easiness in positive selection of $CD8^+$ T cells seems to be in favor of model (ii), presented earlier in the section of positive selection, which assumes that positive selection is not induced by a moderate TCR signal but simply by a strong TCR signal given in thymic cortex.

Allogeneic Transfusion Setting is Preferable

The above method, which is based on the autologous transfusion setting, is bound to face some problems. Not only the production of iPSCs for each patient will be costly, it will also be difficult to ensure TCR affinity quality, and it will need a substantial amount of time (more than 6 months) to prepare regenerated CTLs. We will now discuss how an allogeneic transfusion strategy will solve these problems. In order to make such a strategy work, it will be required to establish a "T-iPSC bank".

In this scheme, T-iPSCs are produced from tumor antigen-specific CTLs collected from healthy donors (Fig. 2.12). Regenerated CTLs will be tested with regard to TCR affinity and ability to give rise to T cells, and if sufficient in quality, such T-iPSCs will be stored as frozen stocks in the bank. Regenerated CTLs themselves will also be frozen and pooled, and they will be immediately used when a HLA-matched patient develops cancer expressing the same tumor antigen. HLA haplotype-homozygous donors are desired as a source of T-iPSCs, since regenerated cells can be given to a patient having the same HLA haplotype allele as a heterozygote. It is also possible to make iPSCs equivalent to T-iPSCs by transducing non-T derived iPSCs with a defined TCR gene. In any cases, it should be tested *in vitro* whether regenerated CTLs happen to exhibit alloreactivity to

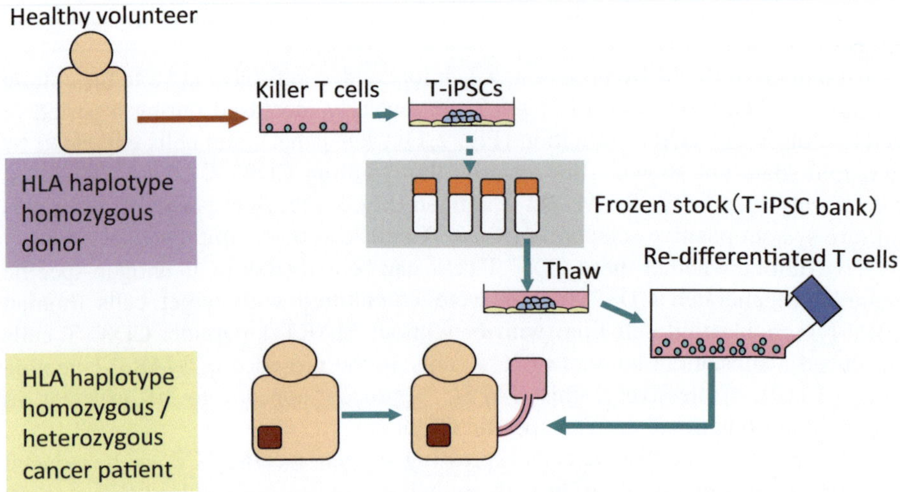

Fig. 2.12 Application of the method to the allogeneic transfusion setting
T-iPSCs are produced from tumor antigen-specific CTLs collected from healthy HLA haplotype-homozygous donors. Regenerated CTLs will be tested with regard to TCR affinity. If they pass quality testing T-iPSCs will be stored as frozen stocks in the T-iPSC bank. Frozen and pooled regenerated CTLs could then be immediately applied to a HLA-matched patient bearing cancer expressing the same tumor antigen

recipient cells, before the use in patients. In addition, there will be another big advantage: patients receiving regenerated CTLs will be free from malignant transformation of transferred CTLs. This is because allogeneic grafts will be eventually rejected based on mismatch of minor histocompatibility antigens even in the HLA-matched case.

Concluding Remarks

In this article, we have presented current knowledge on T cell development and elaborated on the question to what extent T cell development can be recapitulated *in vitro*. Apart from the fact that the studies mentioned in the latter part will be directly applicable in clinical settings, we argue that such *in vitro* studies can be principally regarded as a synthetic biology approach. In other words, when one can control the entire developmental process of a certain type of cell *in vitro* in a culture system using a limited number of defined factors, then it can be said that most of the essential parts regarding the development of that cell type have been clarified. In this respect, in the field of T cell development, we feel that the goal is near. In fact, generation of CD8 T cells starting from ES/iPS cells can be almost perfectly recapitulated *in vitro*, although feeder cells are required for efficient induction. We trust that in the near future it will become possible to induce other types of T cells, and also to do so in a completely defined xeno-free culture system. Thus, we strongly believe that our approach will help clarify essential issues governing the developmental biology of T cells.

References

Adolfsson J, Mansson R, Buza-Vidas N et al (2005) Identification of Flt3$^+$ lympho-myeloid stem cells lacking erythro-megakaryocytic potential a revised road map for adult blood lineage commitment. Cell 121:295–306

Arner E, Daub CO, Vitting-Seerup K et al (2015) Gene regulation. Transcribed enhancers lead waves of coordinated transcription in transitioning mammalian cells. Science 347:1010–1014

Awong G, Herer E, Surh CD et al (2009) Characterization in vitro and engraftment potential in vivo of human progenitor T cells generated from hematopoietic stem cells. Blood 114:972–982

Balciunaite G, Ceredig R, Rolink AG (2005) The earliest subpopulation of mouse thymocytes contains potent T, significant macrophage, and natural killer cell but no B-lymphocyte potential. Blood 105:1930–1936

Bell JJ, Bhandoola A (2008) The earliest thymic progenitors for T cells possess myeloid lineage potential. Nature 452:764–767

Calderon L, Boehm T (2011) Three chemokine receptors cooperatively regulate homing of hematopoietic progenitors to the embryonic mouse thymus. Proc Natl Acad Sci U S A 108:7517–7522

Calderon L, Boehm T (2012) Synergistic, context-dependent, and hierarchical functions of epithelial components in thymic microenvironments. Cell 149:159–172

Doulatov S, Notta F, Eppert K et al (2010) Revised map of the human progenitor hierarchy shows the origin of macrophages and dendritic cells in early lymphoid development. Nat Immunol 11:585–593

Dunon D, Allioli N, Vainio O et al (1999) Quantification of T-cell progenitors during ontogeny: thymus colonization depends on blood delivery of progenitors. Blood 93:2234–2243

Fusaki N, Han H, Nishiyama A et al (2009) Efficient induction of transgene-free human pluripotent stem cells using a vector based on Sendai virus, an RNA virus that does not integrate into the host genome. Proc Jpn Acad Ser B Phys Biol Sci 85:348–362

Gorgens A, Radtke S, Mollmann M et al (2013) Revision of the human hematopoietic tree: granulocyte subtypes derive from distinct hematopoietic lineages. Cell Rep 3:1539–1552

Harman BC, Jenkinson WE, Parnell SM et al (2005) T/B lineage choice occurs prior to intrathymic Notch signaling. Blood 106:886–892

Hodi FS, O'Day SJ, McDermott DF et al (2010) Improved survival with ipilimumab in patients with metastatic melanoma. N Engl J Med 363:711–723

Honey K, Nakagawa T, Peters C et al (2002) Cathepsin L regulates CD4$^+$ T cell selection independently of its effect on invariant chain: a role in the generation of positively selecting peptide ligands. J Exp Med 195:1349–1358

Hozumi K, Mailhos C, Negishi N et al (2008) Delta-like 4 is indispensable in thymic environment specific for T cell development. J Exp Med 205:2507–2513

Ikawa T, Masuda K, Lu M et al (2004a) Identification of the earliest prethymic T-cell progenitors in murine fetal blood. Blood 103:530–537

Ikawa T, Kawamoto H, Wright LY et al (2004b) Long-term cultured E2A-deficient hematopoietic progenitor cells are pluripotent. Immunity 20:349–360

Ikawa T, Hirose S, Masuda K et al (2010) An essential developmental checkpoint for production of the T cell lineage. Science 329:93–96

Ikawa T, Masuda K, Huijskens MJ et al (2015) Induced developmental arrest of early hematopoietic progenitors leads to the generation of leukocyte stem cells. Stem Cell Reports 5:716–727

Itoi M, Kawamoto H, Katsura Y et al (2001) Two distinct steps of immigration of hematopoietic progenitors into the early thymus anlage. Int Immunol 13:1203–1211

Jenkinson EJ, Franchi LL, Kingston R et al (1982) Effect of deoxyguanosine on lymphopoiesis in the developing thymus rudiment in vitro: application in the production of chimeric thymus rudiments. Eur J Immunol 12:583–587

Johnson LA, Morgan RA, Dudley ME et al (2009) Gene therapy with human and mouse T-cell receptors mediates cancer regression and targets normal tissues expressing cognate antigen. Blood 114:535–546

Jotereau F, Heuze F, Salomon-Vie V et al (1987) Cell kinetics in the fetal mouse thymus: precursor cell input, proliferation, and emigration. J Immunol 138:1026–1030

Katsura Y, Kawamoto H (2001) Stepwise lineage restriction of progenitors in lympho-myelopoiesis. Int Rev Immunol 20:1–20

Kawamoto H (2006) A close developmental relationship between the lymphoid and myeloid lineages. Trends Immunol 27:169–175

Kawamoto H, Ohmura K, Katsura Y (1997) Direct evidence for the commitment of hematopoietic stem cells to T, B and myeloid lineages in murine fetal liver. Int Immunol 9:1011–1019

Kawamoto H, Ohmura K, Katsura Y (1998) Presence of progenitors restricted to T, B, or myeloid lineage, but absence of multipotent stem cells, in the murine fetal thymus. J Immunol 161:3799–3802

Kawamoto H, Ikawa T, Ohmura K et al (2000) T cell progenitors emerge earlier than B cell progenitors in the murine fetal liver. Immunity 12:441–450

Kawamoto H, Ikawa T, Masuda K et al (2010) A map for lineage restriction of progenitors during hematopoiesis: the essence of the myeloid-based model. Immunol Rev 238:23–36

Koch U, Fiorini E, Benedito R et al (2008) Delta-like 4 is the essential, nonredundant ligand for Notch1 during thymic T cell lineage commitment. J Exp Med 205:2515–2523

La Motte-Mohs RN, Herer E, Zuniga-Pflucker JC (2005) Induction of T-cell development from human cord blood hematopoietic stem cells by Delta-like 1 in vitro. Blood 105:1431–1439

Li L, Leid M, Rothenberg EV (2010a) An early T cell lineage commitment checkpoint dependent on the transcription factor Bcl11b. Science 329:89–93

Li P, Burke S, Wang J et al (2010b) Reprogramming of T cells to natural killer-like cells upon Bcl11b deletion. Science 329:85–89

Liu C, Saito F, Liu Z et al (2006) Coordination between CCR7- and CCR9-mediated chemokine signals in prevascular fetal thymus colonization. Blood 108:2531–2539

Loh YH, Hartung O, Li H et al (2010) Reprogramming of T cells from human peripheral blood. Cell Stem Cell 7:15–19

Lu M, Kawamoto H, Katsube Y et al (2002) The common myelolymphoid progenitor: a key intermediate stage in hemopoiesis generating T and B cells. J Immunol 169:3519–3525

Lu M, Tayu R, Ikawa T et al (2005) The earliest thymic progenitors in adults are restricted to T, NK, and dendritic cell lineage and have a potential to form more diverse TCRbeta chains than fetal progenitors. J Immunol 175:5848–5856

Luc S, Luis TC, Boukarabila H et al (2012) The earliest thymic T cell progenitors sustain B cell and myeloid lineage potential. Nat Immunol 13:412–419

Masuda K, Itoi M, Amagai T et al (2005a) Thymic anlage is colonized by progenitors restricted to T, NK, and dendritic cell lineages. J Immunol 174:2525–2532

Masuda K, Kubagawa H, Ikawa T et al (2005b) Prethymic T-cell development defined by the expression of paired immunoglobulin-like receptors. EMBO J 24:4052–4060

Masuda K, Kakugawa K, Nakayama T et al (2007) T cell lineage determination precedes the initiation of TCR beta gene rearrangement. J Immunol 179:3699–3706

Meek B, Cloosen S, Borsotti C et al (2010) In vitro-differentiated T/natural killer-cell progenitors derived from human $CD34^+$ cells mature in the thymus. Blood 115:261–264

Morgan RA, Dudley ME, Wunderlich JR et al (2006) Cancer regression in patients after transfer of genetically engineered lymphocytes. Science 314:126–129

Murata S, Sasaki K, Kishimoto T et al (2007) Regulation of $CD8^+$ T cell development by thymus-specific proteasomes. Science 316:1349–1353

Nakano T, Kodama H, Honjo T (1994) Generation of lymphohematopoietic cells from embryonic stem cells in culture. Science 265:1098–1101

Nehls M, Pfeifer D, Schorpp M et al (1994) New member of the winged-helix protein family disrupted in mouse and rat nude mutations. Nature 372:103–107

Nishikawa S, Ogawa M, Nishikawa S et al (1988) B lymphopoiesis on stromal cell clone: stromal cell clones acting on different stages of B cell differentiation. Eur J Immunol 18:1767–1771

Nutt SL, Heavey B, Rolink AG et al (1999) Commitment to the B-lymphoid lineage depends on the transcription factor Pax5. Nature 401:556–562

Ohmura K, Kawamoto H, Fujimoto S et al (1999) Emergence of T, B, and myeloid lineage-committed as well as multipotent hemopoietic progenitors in the aorta-gonad-mesonephros region of day 10 fetuses of the mouse. J Immunol 163:4788–4795

Okamura RM, Sigvardsson M, Galceran J et al (1998) Redundant regulation of T cell differentiation and TCRalpha gene expression by the transcription factors LEF-1 and TCF-1. Immunity 8:11–20

Park IH, Zhao R, West JA et al (2008) Reprogramming of human somatic cells to pluripotency with defined factors. Nature 451:141–146

Peschon JJ, Morrissey PJ, Grabstein KH et al (1994) Early lymphocyte expansion is severely impaired in interleukin 7 receptor-deficient mice. J Exp Med 180:1955–1960

Pongubala JM, Northrup DL, Lancki DW et al (2008) Transcription factor EBF restricts alternative lineage options and promotes B cell fate commitment independently of Pax5. Nat Immunol 9:203–215

Porritt HE, Rumfelt LL, Tabrizifard S et al (2004) Heterogeneity among DN1 prothymocytes reveals multiple progenitors with different capacities to generate T cell and non-T cell lineages. Immunity 20:735–745

Pronk CJ, Rossi DJ, Mansson R et al (2007) Elucidation of the phenotypic, functional, and molecular topography of a myeloerythroid progenitor cell hierarchy. Cell Stem Cell 1:428–442

Pui JC, Allman D, Xu L et al (1999) Notch1 expression in early lymphopoiesis influences B versus T lineage determination. Immunity 11:299–308

Radtke F, Wilson A, Stark G et al (1999) Deficient T cell fate specification in mice with an induced inactivation of Notch1. Immunity 10:547–558

Ramond C, Berthault C, Burlen-Defranoux O et al (2014) Two waves of distinct hematopoietic progenitor cells colonize the fetal thymus. Nat Immunol 15:27–35

Robbins PF, Morgan RA, Feldman SA et al (2011) Tumor regression in patients with metastatic synovial cell sarcoma and melanoma using genetically engineered lymphocytes reactive with NY-ESO-1. J Clin Oncol 29:917–924

Rodewald HR, Ogawa M, Haller C et al (1997) Pro-thymocyte expansion by c-kit and the common cytokine receptor gamma chain is essential for repertoire formation. Immunity 6:265–272

Rosenberg SA (2011) Cell transfer immunotherapy for metastatic solid cancer–what clinicians need to know. Nat Rev Clin Oncol 8:577–585

Rosenberg SA, Packard BS, Aebersold PM et al (1988) Use of tumor-infiltrating lymphocytes and interleukin-2 in the immunotherapy of patients with metastatic melanoma. A preliminary report. N Engl J Med 319:1676–1680

Rossi FM, Corbel SY, Merzaban JS et al (2005) Recruitment of adult thymic progenitors is regulated by P-selectin and its ligand PSGL-1. Nat Immunol 6:626–634

Rothenberg EV, Moore JE, Yui MA (2008) Launching the T-cell-lineage developmental programme. Nat Rev Immunol 8:9–21

Sasaki K, Takada K, Ohte Y et al (2015) Thymoproteasomes produce unique peptide motifs for positive selection of $CD8^+$ T cells. Nat Commun 6:7484

Schmitt TM, Zuniga-Pflucker JC (2002) Induction of T cell development from hematopoietic progenitor cells by delta-like-1 in vitro. Immunity 17:749–756

Seki T, Yuasa S, Oda M et al (2010) Generation of induced pluripotent stem cells from human terminally differentiated circulating T cells. Cell Stem Cell 7:11–14

Takada K, Van Laethem F, Xing Y et al (2015) TCR affinity for thymoproteasome-dependent positively selecting peptides conditions antigen responsiveness in CD8 T cells. Nat Immunol 16:1069–1076

Timmermans F, Velghe I, Vanwalleghem L et al (2009) Generation of T cells from human embryonic stem cell-derived hematopoietic zones. J Immunol 182:6879–6888

Ting CN, Olson MC, Barton KP et al (1996) Transcription factor GATA-3 is required for development of the T-cell lineage. Nature 384:474–478

Topalian SL, Hodi FS, Brahmer JR et al (2012) Safety, activity, and immune correlates of anti-PD-1 antibody in cancer. N Engl J Med 366:2443–2454

Ueno T, Liu C, Nitta T et al (2005) Development of T-lymphocytes in mouse fetal thymus organ culture. Methods Mol Biol 290:117–133

Varnum-Finney B, Brashem-Stein C, Bernstein ID (2003) Combined effects of Notch signaling and cytokines induce a multiple log increase in precursors with lymphoid and myeloid reconstituting ability. Blood 101:1784–1789

Viret C, Lamare C, Guiraud M et al (2011) Thymus-specific serine protease contributes to the diversification of the functional endogenous CD4 T cell receptor repertoire. J Exp Med 208:3–11

Vizcardo R, Masuda K, Yamada D et al (2013) Regeneration of human tumor antigen-specific T cells from iPSCs derived from mature $CD8^{+}$ T cells. Cell Stem Cell 12:31–36

von Freeden-Jeffry U, Vieira P, Lucian LA et al (1995) Lymphopenia in interleukin (IL)-7 gene-deleted mice identifies IL-7 as a nonredundant cytokine. J Exp Med 181:1519–1526

Wada H, Masuda K, Satoh R et al (2008) Adult T-cell progenitors retain myeloid potential. Nature 452:768–772

Weber BN, Chi AW, Chavez A et al (2011) A critical role for TCF-1 in T-lineage specification and differentiation. Nature 476:63–68

Wolchok JD, Kluger H, Callahan MK et al (2013) Nivolumab plus ipilimumab in advanced melanoma. N Engl J Med 369:122–133

Yang S, Liu F, Wang QJ et al (2011) The shedding of CD62L (L-selectin) regulates the acquisition of lytic activity in human tumor reactive T lymphocytes. PLoS One 6, e22560

Yoshida T, Ng SY, Zuniga-Pflucker JC et al (2006) Early hematopoietic lineage restrictions directed by Ikaros. Nat Immunol 7:382–391

Yu S, Zhou X, Steinke FC et al (2012) The TCF-1 and LEF-1 transcription factors have cooperative and opposing roles in T cell development and malignancy. Immunity 37:813–826

Zhang JA, Mortazavi A, Williams BA et al (2012) Dynamic transformations of genome-wide epigenetic marking and transcriptional control establish T cell identity. Cell 149:467–482

Chapter 3
Strategies for Thymic Regeneration: Recent Advances Towards Clinical Therapy

Marco Barsanti, Michael Hun, Richard Boyd, and Ann Chidgey

Abstract The thymus plays a critical role in maintaining immune well-being, but paradoxically undergoes progressive age-related atrophy. Many efforts have been made to deepen our understanding of its biology, with future therapies aimed at re-establishing T cell production in immunocompromised patients. This includes the elderly, and patients undergoing cytoablative treatments or chronically infected with immunotropic viruses. Here we discuss major pre-clinical approaches to thymic regeneration, as a representation of the strategies potentially able to be utilised therapeutically. We outline thymic physiology and development to assist in understanding the rationale behind each regenerative strategy – broadly, reactivation of endogenous thymic epithelial progenitor cells, *de novo* generation of thymic epithelium, and the facilitating effects on these of bioengineering. Continued advancement in these approaches may lead to their clinical translation for the recovery of immune competence in the aged or immune deficient, and tolerance applications in the developing field of stem cell therapeutics.

Abbreviations

allo-HSCT	Allogeneic HSCT
ADE	Anterior definitive endoderm
Aire	Autoimmune regulator
Bcl2	B cell lymphoma 2
biomaterials	Biomimetic materials
BM	Bone marrow
BMT	Bone marrow transplantation
Bmp4	Bone morphogenetic protein 4
CCL21	Chemokine (C-C motif) ligand 21
CCR7	Chemokine (C-C motif) receptor 7
CXCL12	Chemokine (C-X-C motif) ligand 12

M. Barsanti • M. Hun • R. Boyd • A. Chidgey (✉)
Department of Anatomy and Developmental Biology, Monash University, Melbourne, Australia
e-mail: ann.chidgey@monash.edu

T. Watanabe, Y. Takahama (eds.), *Synthetic Immunology*,
DOI 10.1007/978-4-431-56027-2_3

CLP	Common lymphoid progenitor
cTEC	Cortical
CMJ	Cortico-medullary junction
CMJ	Cortico-medullary junction
K8	Cytokeratin-8
Dll	Delta-like ligands
DC	Dendritic cell
DGCR8	DiGeorge syndrome chromosomal region 8
DP	Double positive
(E)11.5	Embryonic day
ESCs	Embryonic stem cells
EAE	Experimental autoimmune encephalomyelitis
ECM	Extracellular matrix
Fgfr2IIIb	Fgf receptor 2 variant IIIb
Fgf8	Fibroblast growth factor 8
Foxn1	Forkhead box-containing gene n1
Foxa2	Forkhead-box gene a2
GPCRs	G protein-coupled receptors
GRL	Ghrelin
Gcm2	Glial cell missing homologue 2
GvHD	Graft versus host disease
GH	Growth hormone
GHRH	Growth hormone releasing hormone
HSCT	Hematopoietic stem cell transplantation
Hox	Homeobox
HIV	Human immunodeficiency virus
HIF-1α	Hypoxia inducible factor-1 alpha
iPSCs	Induced pluripotent stem cells
iTECs	Induced TECs
Igf1	Insulin-like growth factor 1
IFNα	Interferon alpha
IL-7	Interleukin-7
Jag	Jagged
KGF	Keratinocyte growth factor
KO	Knock-out
Klf4	Kruppel-like factor 4
LRCs	Label retaining cells
LHRH	Luteinising hormone releasing hormone
LTi	Lymphoid tissue inducer
MHC	Major histocompatibility complex
mTEC	Medullary
ME-ADE-PE	Mesendoderm to anterior definitive endoderm to pharyngeal endoderm
miRNAs	MicroRNAs

MEFs	Mouse embryonic fibroblasts
MTS24	Mouse thymic stroma 24
mTEChi	mTECs characterised by high MHCII expression
mTEClo	mTECs characterised by low MHCII expression
MafB	Musculoaponeurotic fibrosarcoma oncogene homologue B
MOG	Myelin oligodendrocyte glycoprotein
NC	Neural crest
OKSM	Oct4 Klf4, Sox2 and c-Myc
Oct4	Octamer-binding transcription factor 4
Pax	Paired box gene
Eya	Pax-Eyes absent homologue
Plet-1	Placenta-expressed transcript-1
PSGL1	Platelet (P)-selectin glycoprotein ligand 1
Pdgfrα	Platelet-derived growth factor receptor alpha
PEG	Polyethylene glycol
ROS	Reactive oxygen species
RA	Retinoic acid
RNAi	RNA interference
Sox2	Sex determining region Y-box 2
SSA	Sex steroid ablation
Six	Sine oculis-related homeobox homologue
SP	Single positive
SMC	Smooth muscle cell
SDS	Sodium dodecyl sulfate
SST	Somatostatin
Shh	Sonic hedgehog
Spry	Sprouty
SCF	Stem cell factor
SL-TBI	Sub-lethal total body irradiation
Tbx-1	T box gene
TCR	T cell receptors
Treg	T regulatory
pp3	Third pharyngeal pouch
3D	Three-dimensional
TECs	Thymic epithelial cells
TEPC	Thymic epithelial progenitor cell
TSC	Thymic stromal cells
TRAs	Tissue-restricted self-antigens
TGF-β	Transforming growth factor-β
TA	Transit amplifying
TN	Triple negative
TNF	Tumour necrosis factor
p63	Tumour protein 63
UEA-1	Ulex europaeus agglutinin-1

Vegf	Vascular-endothelial growth factor
c-Myc	v-myc avian myelocytomatosis viral oncogene homolog
WT	Wild-type

Introduction

An Overview of the Thymus

Since antiquity the thymus has remained surrounded by a halo of mystery. The ancient Greeks believed the thymus to house the human soul, and associated its characteristic shape with the thyme leaf. It was noted that this bilobular (in mammals) organ grew to its full extent before puberty, and then began a slow decline toward atrophy. Due to an abundance of resident apoptotic cells and little known role in humans, the thymus was believed to be a vestigial lymphocyte graveyard until the 1960s. In 1961, Jacques Miller unveiled the true function of the thymus in host immunity (Miller 2002). Through young murine thymectomy experiments, he demonstrated this organ to be crucial for the generation of key immune cells – T lymphocytes.

The thymus is the central lymphoid organ responsible for generating T lymphocytes. It attracts circulating hematopoietic lymphoid progenitors, and subsequently generates cohorts of mature, naïve, self-tolerant T cells expressing a broad diversity of antigen specific T cell receptors (TCR) for release back into the bloodstream. This process encompasses a closely connected network of cellular interactions, and multiple key molecules that are expressed on non-hematopoietic thymic stromal cells (TSC) anchored to extracellular matrix (ECM) across a three-dimensional (3D) plane (Boyd et al. 1993; Jenkinson et al. 1992; van Ewijk et al. 1982, 1999). Intimately related components and mechanisms are involved in these complex interactions between resident stromal cells and developing T cells as they migrate in a directed fashion through specific regions of the thymus, including growth factors, transcriptional networks, cell-to-cell contact, chemokines, cytokines, hormones and microRNAs (miRNAs) (reviewed in (Gray et al. 2005; Lynch et al. 2009; Takahama 2006)).

Thymic involution advances with age, resulting in a profound decline in naïve T cell output. This in turn compromises the host immune system's ability to mount effective immune responses against potential pathogens (Gruver et al. 2007; Linton and Dorshkind 2004). Thymic atrophy therefore has important implications in the effects of chemotherapy and other cytoablative treatments on ageing patients, which are heavily reliant on thymic function for reconstitution of a broad T cell receptor repertoire. Regenerative strategies devised to counteract the functional waning of this primary lymphoid organ continue to be investigated for clinical application (Chidgey et al. 2008).

The Thymus and T Cell Development

Tightly coordinated signals and complex interactions occur between developing T cells and components of the thymic stroma, which comprises predominantly of thymic epithelial cells (TECs). These resident epithelial cells can be divided broadly into inner medullary (mTEC) and outer cortical (cTEC) subgroups. Following entry at the cortico-medullary junction (CMJ) of the thymus, bone marrow (BM) derived lymphoid progenitor cells migrate in a directed fashion through the cortex and ultimately through the medulla, concomitantly undergoing transcriptional and phenotypic changes correlating with stage-specific molecular signatures and surface markers (Blackburn and Manley 2004). The unique three-dimensional lattice that permits thymocyte migration with intimate TSC contact is formed by cortical and medullary epithelia interspersed with thymic fibroblasts and vasculature (van Ewijk et al. 1994, 1999, 2000). Other non-epithelial stromal elements, such as dendritic cells and macrophages, are also integrated into this meshwork (Boyd et al. 1993). Together, the TSCs create the essential microenvironments required to direct and regulate T cell development (Petrie and Zuniga-Pflucker 2007).

Hematopoietic stem cell derived lymphoid lineage-restricted progenitor cells arise in the foetal liver during embryogenesis and in the BM of adults (Weissman and Shizuru 2008). While it is unclear which specific progenitor subpopulations enter the thymus, most seem to be lymphoid lineage-committed, with TEC interaction initiating T cell fate restriction (reviewed in (Dudakov et al. 2010; Zlotoff and Bhandoola 2011)).

During embryogenesis, thymocyte progenitors begin their migration through the thymic capsule and into the avascular thymus at embryonic day (E)11.5 in mice, orchestrated by TEC-derived chemokines such as chemokine (C-C motif) ligand 21 (CCL21), CCL25 and chemokine (C-X-C motif) ligand 12 (CXCL12) (Blackburn and Manley 2004; Holländer et al. 2006; Kawakami et al. 1999; Liu et al. 2005). These signalling proteins regulate thymocyte trafficking by interacting with G protein-coupled receptors (GPCRs) (Annunziato et al. 2001). Only CXCL12 is known to be non-essential for foetal thymic colonisation, with deficiencies in either CXCL12 or its receptor CXCR4 causing no adverse effects on T cell progenitor recruitment (Ara et al. 2003). In contrast, neutralising antibodies targeting CCL21 and CCL25 caused a significant fall in total thymocyte attraction (Liu et al. 2005). These results together indicated CCL21 and CCL25, with their respective receptors chemokine (C-C motif) receptor 7 (CCR7) and CCR9, as critical players in initial thymic colonisation. Once colonised, the immature thymocyte developmental pathway triggers the differentiation and expansion of mature cTECs from thymic epithelial progenitor cells (Hollander et al. 1995; van Ewijk et al. 2000).

TEC-thymocyte cross-talk is crucial in fostering terminal epithelial cell differentiation, marked by Major Histocompatibility Complex (MHC)-II expression in cTECs and mTECs, and by the transcription factor, autoimmune regulator (Aire) in mTECs that express high levels of MHCII (MHCII^{hi} Ulex europaeus agglutinin-1 (UEA-1)$^{+}$). The Notch signalling pathway plays a major role in this process, with its first three receptors found to be expressed in the epithelium along with their five ligands: Jagged (Jag) 1 and 2, and Delta-like ligands (Dll) 1, 3 and 4 (Masuda et al. 2009). These were shown to be required for the newly formed thymus to support normal early thymocyte differentiation. Furthermore, tumour protein 63 (p63), an important onco-suppressor gene involved in Notch signalling, was found to be highly expressed in both the cTECs and mTECs of the developing thymus (Candi et al. 2007; Kikuchi et al. 2004). Deletion of its isoforms and, in particular, the N-terminal truncated form ΔNp63 were linked to a hypoplastic epithelium and the loss of *Aire* expression through a marked rise in cell apoptosis.

Complete vascularisation of the foetal thymus is ultimately achieved between E14.5 and E15.5, henceforth permitting direct T cell progenitor entry. This marks the migration of a second wave of lymphoid progenitors into the thymus, with seeding of the organ occurring through its newly formed blood vessels (Bleul et al. 2006; Holländer et al. 2006). During this time, the thymic medulla, part of which acquires an Aire^{+} phenotype, starts to differentiate in order to accommodate the newly generated, first-wave derived, CD4 and CD8 single positive (SP) T cells that need to undergo negative selection (Yano et al. 2008). Thymic cellularity hence continues to rise while the organ enlarges to adjust for the continuous flow of hematopoietic progenitors.

Postnatal circulating lymphoid progenitors are continuously attracted into the thymus. These progenitors migrate into the thymic parenchyma via the vasculature at the cortico-medullary junction (CMJ) with P-selectin expressed on thymic endoderm playing a role in attracting the circulating lymphoid progenitor cells expressing platelet (P)-selectin glycoprotein ligand 1 (PSGL1) (Rossi et al. 2005). Here they commit to a T cell lineage by signalling through Notch-Delta ligand interactions with cTEC, and in combination with interleukin-7 (IL-7) signalling, are induced to proliferate, generating a vast pool of immature progeny (Petrie and van Ewijk 2002). The TCR^{-}CD4^{-}CD8^{-} triple negative (TN) cells advance through four major TN phases according to their CD44 and CD25 expression, before they reach a TCR^{+}CD4^{+}CD8^{+} double positive (DP) state (Godfrey and Zlotnik 1993; Godfrey et al. 1993). Expression of chemokine receptors CXCR4, CCR7 and CCR9 drive the outward migration seen during this process (Takahama 2006). Newly generated CD4^{+}CD8^{+} DP thymocytes in the thymic cortex interact with MHC-I or MHC-II molecules that are expressed on the surface of cortical thymic epithelial cells (Goldrath and Bevan 1999; Takahama 2006). DP thymocytes that engage self-peptide bound to MHC through their TCR with low avidity are rescued from apoptotic death by neglect and receive further differentiation signals, collectively termed positive selection; those that bind with high affinity, thereby representing auto-reactive cells, are deleted by apoptosis. Only 3–5 % of developing thymocytes survive this stage (Goldrath and Bevan 1999).

DP thymocytes differentiate into CD4 or CD8 SP cells and relocate to the CCL19 and CCL21 expressing medulla via CCR7-mediated chemotaxis (Kurobe et al. 2006). As CD4 and CD8 SP T cells migrate towards the medulla, high affinity interactions with dendritic cells and mTECs induce apoptosis, a negative selection process known as central tolerance (reviewed in (Blackburn and Manley 2004)). While negative selection also transpires in the thymic cortex, it is the functionally mature CD80^{hi} mTECs that contribute greatest to central tolerance, expressing a vast diversity of genes which represent every tissue (Derbinski et al. 2005). Most of these promiscuously expressed genes are induced by Aire, which acts stochastically through inter-chromosomal activation of gene clusters within specific mTEC subsets (Derbinski et al. 2005; Meredith et al. 2015; Takahama 2006). In concert with Aire, tissue-restricted self-antigens (TRAs) expressed by a subpopulation of mTECs also contribute to establish central tolerance (Brennecke et al. 2015). This is achieved by an orchestrated process that utilises chromatin remodelling to broaden the breadth of tolerising antigens available for presentation to developing thymocytes.

Studies show that S1P-mediated chemotaxis is essential for T cell egress into the periphery, with S1P1 deletion in T cell-specific S1P1 knock-out mice causing an accumulation of mature SP thymocytes in the adult thymus (Allende et al. 2004). Interactions between CCL19 and its receptor CCR7 also play a part in thymocyte emigration, however this role is seen only in neonates (Ueno et al. 2002).

Thymus Organogenesis

Following germ-layer specification, the anterior definitive endoderm (ADE) gives rise to the salivary glands, thyroid, parathyroids, thymus, lungs, stomach, liver and pancreas (Gilbert 2010). It is from one of the segments of the pharynx that the thymus proper arises. Patterning begins after four pharyngeal arches and three pharyngeal pouches have formed. The pharyngeal pouches are bilateral outpocketings of epithelial endoderm, and their patterning is regulated through a gradient of clustered homeobox (*Hox*) genes (Manley and Capecchi 1995, 1998). One gene in particular, *Hoxa3*, was shown to be crucial for correct third pharyngeal pouch (pp3) positioning and the migration of its derivatives.

The preliminary stages of thymic organogenesis occur in unison with another organ that develops from the pp3, the parathyroid. The initial signals that orchestrate pp3 endoderm differentiation include the expression of paired box gene (*Pax*)-*1*, *Pax-9*, fibroblast growth factor 8 *(Fgf8)*, T box gene (*Tbx*)-*1*, the *Pax*-Eyes absent homologue (*Eya*)-Sine oculis-related homeobox homologue (*Six*) cascade, sonic hedgehog (*Shh*) – which is required for parathyroid domain specification as early as E9.5 – and forkhead-box gene a2 *(Foxa2)* (Nowell et al. 2007a; Zamisch et al. 2005). Furthermore, signals from the surrounding neural crest (NC)-derived mesenchyme and pharyngeal cleft ectoderm were demonstrated to be vital for correct pp3 endoderm differentiation, especially during the first stages of its commitment (Blackburn and Manley 2004). For instance, at E10.5, the surrounding mesenchyme was shown to express genes such as *Hoxa3*, *Eya1* and *Six1*, which initiate the patterning of the pouch into two distinct domains – the ventral thymic rudiment and the dorsal parathyroid primordium (Nowell et al. 2007a; Rodewald 2008). Expression of vascular-endothelial growth factor (*Vegf*), *Fgf8* and *Fgf10* was also noted. During this stage, the pharyngeal cleft ectoderm instead expresses *Tbx1* and *Fgf10*, while the endoderm itself begins expressing *Fgf10, Vegf,* and Fgf receptor 2, variant IIIb (*Fgfr2IIIb*). Retinoic acid (RA)-induced patterning was also shown to be involved in inducting pp3 differentiation (Holländer et al. 2006).

At E10.5, bone morphogenetic protein 4 (Bmp4), a member of the transforming growth factor-β (TGF-β) family, was shown to participate in epithelial-mesenchymal interactions (Gordon et al. 2010). Prominent *Bmp4* expression was observed in both the epithelium and the surrounding NC-derived mesenchyme. Although introduction of a *Foxn1::Noggin* transgene – with Noggin being a strong inhibitor of Bmp4 – demonstrated no effect on normal thymocyte development, the stromal component to the thymus was reported to exhibit strong developmental delay coupled with marked hypoplasia and inhibition of normal organ migration (Bleul and Boehm 2005). Furthermore, Bmp4, which is secreted into the ECM, was found to be necessary for thymic lumen closure, initial lymphocyte colonization, separation of the organ from the pharynx, capsule formation, and ventro-caudal organ migration (Patel et al. 2006).

The differentiating pp3 endoderm begins to bud at E11.25, and two distinct domains are evident – a glial cell missing homologue 2 (Gcm2) domain marking the developing parathyroids, and a forkhead box-containing gene n1 (Foxn1) domain, which defines the early thymus. Bmp4 inhibitors such as Chordin and Noggin have also been implicated in the formation of the thymic rudiment. Chordin expression persists until at least E15.5 and is essential for *Tbx1* expression, while Noggin is present in the pharynx by E9.5, marking the parathyroid primordium one day later (Bachiller et al. 2003; Holländer et al. 2006; Patel et al. 2006; Scott et al. 2000). During this stage of thymus development, the trophic action of the surrounding mesenchyme is still required, with its absence resulting in expression inhibition of insulin-like growth factor 1 and 2 (*Igf1* and *2*) early on, and *Fgf7* and *Fgf10* at later stages (Jenkinson et al. 2007). This growth factor-releasing layer is marked by high levels of platelet-derived growth factor receptor alpha (*Pdgfrα*) expression. Although Fgf signalling plays a major part in correcting thymic development, the levels of such mitogens must be tightly regulated at all times. In fact, Gardiner *et al.* (2012) have highlighted the important roles of Sprouty (Spry) 1 and 2, which are Fgf feedback antagonists present in the early thymic anlage that inhibit Fgf signalling in pp3 (Gardiner et al. 2012). Their absence causes Fgf8 hyper-responsiveness that in turn leads to thymic hypoplasia. From E11.5, a leucine-zipper transcription factor, musculoaponeurotic fibrosarcoma oncogene homologue B (*MafB*) is strongly detected in mesenchyme, and shown to be related to the expression of chemotactic signals such as CCL21 and CCL25, which are designed to attract developing hematopoietic precursors (Sultana et al. 2009).

At E12.5, the parathyroids and thymus acquire distinct identities, with the two thymic lobes initiating migration caudally and medially towards the heart, where they will eventually fuse. TECs begin differentiation, firstly in a thymocyte-independent manner through *Foxn1* up-regulation and subsequently via crosstalk, based on complex cell-to-cell interactions between TECs and early lymphoid precursors. During this timepoint, the majority of TECs are aligned to a cortical lineage phenotype, expressing both the proteasome subunit β5t and CD205, suggesting a serial progression from cTEC to mTEC during embryonic thymus development (Alves et al. 2014). Furthermore, most of these E12.5 fetal TECs appear to have bipotent progenitor capacity as demonstrated by their ability to produce both mature cortical and medullary epithelium when grafted under the kidney capsule of immunodeficient *nude* mice as a purified population or as single cells marked with enhanced yellow fluorescent protein (Bennett et al. 2002; Flanagan 1966; Rossi et al. 2006).

By E15.5, TEC differentiation is well underway. It is at this stage that Gill *et al.* (2002) identified a thymic epithelial progenitor cell (TEPC) population using the mouse thymic stroma 24 (MTS24) antibody, which binds placenta-expressed transcript-1 (Plet-1) (Depreter et al. 2008; Gill et al. 2002). These $MTS24^{+}$ cells express both cytokeratin-8 (K8; cTEC marker) and cytokeratin-5 (K5; mTEC marker) and their functional bipotent potential was demonstrated by the generation of both cTEC and mTEC when only 2500 E15.5 $MTS24^{+}$ cells alone were grafted under the kidney capsule of nude mice (Gill et al. 2002). However, in the adult, only a small subset of $MTS24^{+}$ thymic epithelial cells remain and these are found mostly in the medulla. This phenotypic change coincided with an alteration in their functional potential as progenitors (Rossi et al. 2007a; Wong et al. 2014).

*Forkhead-Box Transcription Factor n1 (*Foxn1*)*

Among the transcription factors that play an important role in thymus organogenesis and homeostasis, Foxn1 is indisputably the most crucial (Nowell et al. 2007a; Romano et al. 2013). Mutation or deletion of this gene alone results in severe thymic aplasia in the former, and total athymia in the latter case, with a significant decrease in T cell output that results in profound immune deficiency (Nehls et al. 1994, 1996). The manifestation of extensive alopecia due to abnormal keratinisation of the hair shaft, which creates the "nude" phenotype, highlights the importance of this gene in other tissues, such as skin epithelium (Johns et al. 2005; Petersen et al. 2010). A study revealed stochastic induction of *Foxn1* expression in cells of the "nude" mouse anlage to promote both cTEC and mTEC differentiation, hence confirming Foxn1 to be necessary for TEPC differentiation (Bleul et al. 2006).

Different levels of *Foxn1* mRNA are required at various nodal points of cell fate decision programs towards mature TEC phenotypes, and for the expression of chemotactic, T cell progenitor-recruiting chemokines such as CCL25 (Blackburn et al. 1996; Liu et al. 2006). Unlike lymphoid progenitors, TEPCs can be found in the "nude" thymic rudiment and are established in the pp3 before the expression of *Foxn1* at E11.25 (Itoi et al. 2001). Although the initial stages of TEPC differentiation are carried out in the absence of Foxn1, TEPC expansion incorporating T cell crosstalk dependent steps that follow do require its expression. The molecular dissection of *Foxn1* by Su et al. (2003) provides strong evidence of the role of a hypomorphic version of its allele (*Foxn1*$^{\Delta}$, lacking a functional exon 3) in fostering advanced, crosstalk-induced TEC progression towards their mature stages (Su et al. 2003). Furthermore, there are reports suggesting that Foxn1 may have a role in balancing differentiation and proliferation in TECs, similar to the role established in keratinocytes (Baxter and Brissette 2002; Brissette et al. 1996; Itoi et al. 2001). Analysis of the nude thymic rudiment reveals that it closely resembles lung epithelium, which suggests that Foxn1 might also be involved in early anterior endoderm specification (Dooley et al. 2005).

Unlike the well-documented contribution of Foxn1 in thymus development, the role it covers in adult homeostasis is less understood, with at times contradictory findings (Chen et al. 2009). The winged helix transcription factor is expressed at various levels in all TEC subsets after birth, and may be involved in the differentiation of immature mTECs – characterised by low MHCII expression (mTEC^{lo}) – towards their mature counterpart, which is characterised by high MHCII expression (mTEC^{hi}). Although the aforementioned study implicates mTEC^{hi} as the subset expressing the highest levels of *Foxn1*, others have reported cTEC^{hi} to have the most prominent *Foxn1* expression (Nowell et al. 2011).

Thymus Ageing

Involution of the thymic epithelium can occur as early as the first year of life in humans, but becomes more pronounced from later childhood to early teenage years, with expansion of the perivascular space being the most prominent histological change (Steinmann et al. 1985). This adipose-rich space replaces the true thymic epithelial volume over time, estimated at approximately 3 % per year (Gruver et al. 2007). Naïve T cell output falls as a consequence of thymus functional decline, as does the ratio between naïve T cells and memory T cells residing in the periphery (Aspinall and Andrew 2000; Gruver et al. 2007). As a consequence, the efficiency and effectiveness of the host immune response decreases with age, leaving individuals with a greater disease susceptibility (Aspinall and Andrew 2000).

A gradual age-related decrease at both the mRNA and protein levels implicated Foxn1 in this thymic involution, as did findings that controlled reduction of Foxn1 levels accelerated atrophy (Cheng et al. 2010; Guo et al. 2012; Itoi et al. 2007; Ortman et al. 2002; Sun et al. 2010). Further supporting the role of Foxn1 in this process, addition of *Foxn1* cDNA appeared to reverse involution, while its overexpression alleviated the aged phenotype in mice (Sun et al. 2010; Zook et al. 2011). Moreover, stromal cell profiling has recently revealed the transcriptional changes that occur during early thymic involution (Ki et al. 2014). Such findings have led to the discovery of an unanticipated biological function for mTEChi cells. That is, in addition to their canonical role in negative selection, these cells exhibit a putative role against infection, highlighted by their expression of early innate immune response-related genes.

The progressive decline in MHC class II^+ TEC production of IL-7, a cytokine with an important role in thymopoiesis and peripheral T cell expansion, has also been linked to thymic involution (Andrew and Aspinall 2002; Henson et al. 2005; Mackall et al. 2001; Steinmann et al. 1985). It is well established that several stages of thymocyte development, ranging from the DN precursor phases through to memory T cell formation, require signalling between IL-7 and its receptor IL-7R (van den Brink et al. 2004). Furthermore, IL-7 regulates T cell homeostasis in the periphery, efficaciously acts as a vaccine adjuvant, and exerts an anti-apoptotic effect on lymphocytes (Unsinger et al. 2010; Schluns et al. 2000; Melchionda et al. 2005).

In addition to IL-7, other key molecules and biochemical pathways have been identified as major players in thymic atrophy. Of note is the decline in growth hormone (GH) production with age, and the negative effects of sex steroids on thymic functionality and T cell output (Hirokawa et al. 2015; Ventevogel and Sempowski 2013). Strikingly, targeted destruction of the anterior hypothalamic area of the brain increased thymic weight and was related to increased serum levels of GH (Utsuyama et al. 1997). Decreasing levels of growth hormone releasing hormone (GHRH), a promoter of GH production, and increased somatostatin (SST), its counterpart, were also found in aged mice (Hirokawa et al. 2001). From their collated findings, Hirokawa and colleagues proposed that the progressive post-natal decrease in GH and consequent decline in immune function act to ensure greater life span by reducing Igf1 production (Hirokawa et al. 2015).

Another contributing factor to the decreased naïve T cell output could be supply of BM derived progenitors. The BM is responsible for creating lymphoid progenitors that will either differentiate towards the B lineage *in situ*, or exported to the thymus via the bloodstream for the production of naïve, self-tolerant T cells. The paradoxical increase in BM residing HSCs with ageing is now understood to be a hallmark of impaired quiescence and subsequent decline in the differentiation capacity of these cells (Chambers and Goodell 2007; de Haan and Van Zant 1997). A skewing towards myelopoiesis is also observed (Sudo et al. 2000). This decline in common lymphoid progenitor (CLP) production, which is partially explained by increased reactive oxygen species (ROS)-induced DNA damage, may represent the first step in a cascade of events that dramatically decreases the efficacy of thymic seeding by lymphoid precursors (Chambers and Goodell 2007). However, transplantation of various aged thymi into newborn nude recipients – which would provide young BM progenitors – revealed a reduction in T cell response with thymus age (Hirokawa et al. 1982).

The aged endosteal niche contains an increased adipocyte/osteoblast ratio (Kassem and Marie 2011; Tuljapurkar et al. 2011). This increase appears to contribute to a progressive disruption of the hypoxic environment needed by HSCs for maintenance of their regenerative capacities, with sustained VEGFα and hypoxia inducible factor-1 alpha (HIF-1α) production needed for preserving their mainly glycolytic, low ROS-producing metabolism (Chinn et al. 2012).

Similar oxidative stress-induced, metabolic damage has been observed in the aged thymus; transcriptome analysis of the atrophic thymus identified mitochondrial disruption induced by catalase deficiency (Griffith et al. 2015). This enzyme is essential for hydrogen peroxide detoxification and reduction in toxicities produced through mitochondrial ROS, and was found to be particularly prominent in cortical cells. Using heterochronic parabiosis, young circulating factors were not sufficient to drive regeneration of the aged thymus (Seach et al. 2012) also supporting TEC-intrinsic changes during ageing. Taken together, these findings demonstrate the importance of understanding the mechanisms behind age-related thymic atrophy, in order to implement effective regenerative strategies.

The Clinical Impact of Thymus Ageing and Strategies for Its Regeneration to Replenish the Peripheral T Cell Pool

High levels of morbidity remain one of the most challenging clinical problems facing cancer patients undergoing high dose cytoablative therapy (van den Brink et al. 2004). Most problematic in adults/elderly, it is directly aligned with thymus atrophy, with chemotherapy and radiation treatments further damaging the residual ageing thymus and compounding its atrophy (Mackall et al. 1997; Rossi et al. 2007b). Prompt recovery of the immune system following cytoablation is required to reduce the risk of opportunistic infections, chronic inflammation and activation of latent infections. However, in the elderly, reconstitution of T cells following chemotherapy and other cytoablative treatments heavily relies on peripheral T cell expansion leading to a peripheral T cell pool containing only dominant memory T cell clones. Repair and regeneration of thymic function is required to generate new naïve T cells to reconstitute peripheral T cell diversity (Nowell et al. 2007b; Rossi et al. 2007b).

Pre-clinical approaches to restore thymic function to existing residual or damaged thymic tissue, include sex steroid ablation (SSA), administration of cytokines and growth factors such as IL-7, IL-22 or keratinocyte growth factor (KGF); hormone supplementation such as GH or Ghrelin (GRL) and small RNAs – illustrated in Fig. 3.1 (Boyd et al. 2013; Holländer et al. 2010; Ventevogel and Sempowski 2013). Bio-engineering strategies have been investigated across many different regenerative fields to enhance activation of resident organ specific stem cells, including more recently, the thymus (Fan et al. 2015; Palamaro et al. 2013; Tajima et al. 2015).

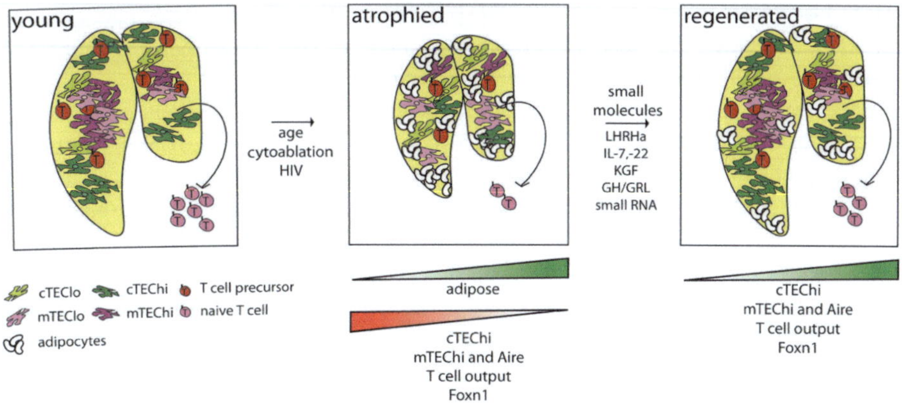

Fig. 3.1 Endogenous re-activation of residual thymic tissue. Thymic damage from ageing, cytoablative therapy or immunotropic viruses results in stromal atrophy and adipocyte infiltration, which decreases cTEChi and Aire^{+} mTEChi subsets, as well as T cell output. Reversal of these effects may be achieved by administration of small molecules such as LHRH antagonists, IL-7, IL-22, KGF, GH/GRL and small RNAs. LHRH antagonists act by inhibiting sex steroid production to transiently induce an expansion in stromal and thymocyte cellularity. Conversely, IL-7 stimulates thymopoiesis, while IL-22 boosts TEC survival and proliferation. In addition to up-regulating IL-7, KGF also promotes temporary TEC proliferation and differentiation. Both GRL and GH enhance overall immunity and restore thymic cellularity by exerting an anti-inflammatory effect, while small RNAs may inhibit key molecules involved in thymic involution. *Aire* autoimmune regulator, *cTEClo* cortical thymic epithelial expressing low MHCII, *cTEChi* cTEC expressing high MHCII, *Foxn1* Forkhead box-containing gene n1, *GH* growth hormone, *GRL* ghrelin, *HIV* human immunodeficiency virus, *IL-7* interleukin 7, *KGF* keratinocyte growth factor, *LHRHa* luteinizing hormone releasing hormone antagonist, *MHCII* major histocompatibility complex class II, *mTEClo* medullary TEC expressing low MHCII, *mTEChi* mTEC expressing high MHCII

Reactivation of the Endogenous Thymus

Sex Steroid Ablation

Serum sex steroid levels surge during adolescence, adversely affecting thymic weight and architecture. Rodent studies confirmed this association between the onset of puberty and thymic involution by demonstrating that castration prior to sexual maturity resulted in delayed atrophy (Greenstein et al. 1986; Greenstein et al. 1987; Heng et al. 2005; Kendall et al. 1990; Olsen et al. 2001; van den Brink et al. 2004; Windmill et al. 1993; Windmill and Lee 1999). In 1987, Greenstein further demonstrated that the more clinically relevant luteinising hormone releasing hormone (LHRH) analogue, as a form of chemical castration, could also induce thymus regeneration (Greenstein et al. 1987). This work highlighted the possibility that androgen suppression therapies could potentially restore naïve T cell output.

Further pre-clinical studies were performed to determine whether castration could regenerate the thymus following damaging cytoablative regimes such as chemotherapy and irradiation associated with bone marrow transplantation

(BMT) or hematopoietic stem cell transplantation (HSCT) and other cancer treatments. Indeed SSA was found to enhance immune recovery following cytoablative treatment, stimulating thymocyte proliferation and differentiation in young and aged, recovery of lymphopoiesis in the BM, and enhanced engraftment and immune reconstitution following autologous and allogeneic HSCT (Dudakov et al. 2009a, b; Goldberg et al. 2005, 2007, 2010; Heng et al. 2005, 2012).

Subsequently, a phase II clinical trial administering a LHRH agonist after myeloablative chemotherapy in BMT patients indicated that naïve $CD4^{+}$ T cell output was enhanced (Sutherland et al. 2008). SSA of aged mice could also restore virus-specific T cells and induce protective effects on tumour induction, however, LHRH agonist treatment of melanoma patients did not increase the frequency of tumour specific CD4 and CD8 T cells after peptide vaccination (Heng et al. 2012; Vence et al. 2013).

Taking into consideration evidence suggesting that TEC loss begins during early post-natal stages, and androgen levels decline with ageing with no direct evidence of associated T cell regeneration, the cause-effect relationship between the two events becomes less clear (Morley 2003). In support of this, it was found that the thymus regenerative effects have only a temporary effect in surgically castrated mice despite permanent loss of sex steroid production (Griffith et al. 2012). Another layer of complexity is evident in the sexual dimorphism of the immune system, and the intricate, often dualistic interactions between oestrogens and thymus in females (Okuyama et al. 1992; Pennell et al. 2012). The effects of SSA in females have not yet been extensively studied, although in the LHRH agonist clinical trial, females responded as well as males (Sutherland et al. 2008). All in all, there is evidence regarding the potential of androgen suppression therapies as a strategy for immune regeneration, but further investigation into new therapeutics with fewer side effects is required. Nevertheless, even a temporary resurgence of naïve T cell generation may be enough to replenish the depleted T cell pool following anti-neoplastic treatments.

Cytokine and Growth Factor Supplementation

It is no mystery that cytokines and growth factors play a major role in thymus and T cell development. They are involved in biological processes ranging from the migration of hematopoietic progenitors, to thymocyte proliferation, epithelial differentiation, fibroblast-TEC interactions, and thymocyte-TEC cross-talk (Ventevogel and Sempowski 2013). Understandably, some growth factors have thus been identified as suitable candidates to replenish these signalling pathways, in an attempt to mirror the pre-pubescent thymic environment in an aged model, to reverse atrophy.

The age related decline in production of the IL-7 glycoprotein as mentioned above, made this interleukin an obvious candidate for immune regeneration. When utilised in metastatic melanoma patients, IL-7 caused a temporary surge in SP T cell number and incremented TCR diversity when administered in HSCT patients

(Perales et al. 2012; Rosenberg et al. 2006). Whilst many studies have confirmed the beneficial effects of IL-7 on thymocyte development, treatments which incorporate this cytokine are not without their concerns (Henson et al. 2005; van den Brink et al. 2004). High doses of IL-7 have been linked to osteoporosis, hyperproliferation, haemolytic anaemia, and other severe side effects, hence it is clear that patient drug sensitivity must be considered prior to therapeutic administration (Henson et al. 2005). CCR9/IL-7 fusion proteins, which were designed to combat such dosage issues, have yielded encouraging results in mice, but whether human thymic function can be significantly enhanced using exogenous IL-7 still remains controversial (Henson et al. 2005; Chu et al. 2004; Storek et al. 2003). Furthermore, this cytokine displayed important limitations as it mainly acts on the T cell pool while leaving thymus size unchanged and TECs mostly unaffected (Sportès et al. 2008).

More recently, the cytokine IL-22 has been brought to attention for its marked regenerative potential after insult. Dudakov et al. (2012) demonstrated that IL-22 can speed thymic recovery after sub-lethal total body irradiation (SL-TBI) in terms of both thymocytes and TEC subsets (Dudakov et al. 2012). They also demonstrated IL-22 receptor expression on TECs, which signals through IL-22 binding to boost T cell survival and proliferation. This signal was revealed to be dependent on lymphoid tissue inducer (LTi)-stimulating dendritic cell (DC) production of IL-23. Interestingly, post-radiation immune recovery was also impaired in IL-22 knockouts (IL-$22^{-/-}$) in comparison to wild-type (WT) mice. Given its relative novelty, more evidence is required in order to determine the full extent of IL-22 function, and whether it can be applied to different atrophy models such as age-related thymic involution.

KGF has also been demonstrated to have regenerative effects on the ageing thymic microenvironment (Min et al. 2007; Alpdogan et al. 2006). This protein belongs to the FGF family and not only plays a crucial role during pp3 commitment and thymus organogenesis, but also regulates both TEC differentiation and proliferation (Rossi et al. 2002, 2007b). By up-regulation of *IL-7* transcription, it appears to exert a stimulatory effect on T cell production (Min et al. 2002). As the effect of KGF is only temporary, repeat treatment courses are required bimonthly (Min et al. 2007). There is contradicting evidence on the effects of KGF with regards to incidence of Graft versus Host Disease (GvHD) in BMT patients (Panoskaltsis-Mortari et al. 1998, 2000; Blazar et al. 2006). Furthermore, a recent study demonstrated an ameliorating effect on immune deficiency with post-allogeneic HSCT (allo-HSCT) in TBI-, but not in chemotherapy-induced HSC depletion (Goldberg et al. 2013). Clinical administration of Palifermin, the pharmaceutical equivalent of KGF, resulted in side effects such as skin irritation (Blazar et al. 2006), possibly due to its involvement in skin keratinocyte differentiation and proliferation (Brauchle et al. 1995). These discrepancies indicate that KGF still requires further investigation to overcome its short-term efficacy and ill-defined influence on the immune system.

Hormone Therapy

GH and its secretagogue and hunger-inducing hormone, Ghrelin, appear to exert regenerative and enhancing effects on the thymus and general immunity (Taub et al. 2010). GH mainly acts by stimulating the hepatocyte-derived secretion of Igf1, and confers apoptosis resistance and increased survival to lymphoid cells (Dorshkind and Horseman 2000; Redelman et al. 2008). GH receptors can be found on almost any cell of both the innate and the adaptive immune system. GH itself is expressed by the thymus, spleen, lymphocyte, and tonsils, suggesting a diverse, paracrine role for this hormone, which is mainly known for its endocrine, musculoskeletal and metabolic function (Hattori 2009). Extensive studies on its thymic effects demonstrated increased TEC proliferation, improved thymic cellularity, enhanced cytokine secretion, as well as augmented ECM production (de Mello Coelho et al. 2002; Savino et al. 2002; Smaniotto et al. 2010). In addition, GH has been demonstrated to elicit CXCL12 levels in TECs, therefore increasing their chemotactic signals toward incoming hematopoietic progenitors (Smaniotto et al. 2005).

GH research has, however, not always delivered positive or unequivocal results. Administration of GH by Rapaport et al. (1986) demonstrated no substantial change in T cell output, antibody prevalence in serum, or innate immune cell activity (Rapaport et al. 1986). A marked reduction in circulating B lymphocytes was noted in this study, alongside an increasing insensitivity to mitogen stimulation and a decreasing CD4 T/T regulatory (Treg) cell ratio. Whilst its role in lymphopoiesis was questioned by the lack of an overt loss of function phenotype in GH-deficient mice, there were contrasting observations, possibly due to differences in housing-related stress (Dorshkind and Horseman 2000; Dorshkind et al. 2003).

Nevertheless, randomized trials investigating a possible role for GH in rescuing HIV-induced thymic atrophy and consequent decreased CD4 T cell output have yielded encouraging results. Consistent and long term increases in thymic size and cellularity were documented, with a 30 % increase in T helper cell production up to one year after discontinuation of therapy (Napolitano et al. 2002, 2008). The main side effects included musculoskeletal pain and the risk of diabetes through the elevated IGF-1.

Significant discoveries have also been made on molecules upstream of GH production (Tannenbaum and Bowers 2001). In particular, GRL was identified as a stimulant for GH release upon binding to a specific receptor expressed by pituitary cells. Further investigations implicated GRL in hunger regulation, adipose tissue levels, and insulin sensitivity when expressed by stomach fundus cells via leptin, a complex interaction with another hunger- and metabolism-regulating hormone (Smith et al. 2005). GRL is produced by the spleen, thymus, lymph nodes and various leukocytes (Dixit et al. 2004; Gnanapavan et al. 2002). Thymi of mice deficient for this hormone or for its receptor displayed greater detrimental effects in terms of age-related TEC damage and T cell development (Dixit et al. 2007).

An acylated derivative of GRL was found to possess anti-inflammatory properties by synergistically inhibiting pro-inflammatory cytokines (IL-1β, IL-6, tumour necrosis factor α (TNFα)), while stimulating immunosuppressive cytokine production (IL-10) (Baatar et al. 2011). Like GH levels, acyl-modified GRL levels significantly drop with ageing, with their replenishment driving immune regeneration by increasing thymic cellularity as well as enhancing BMT engraftment (Dixit et al. 2007; Taub et al. 2010; Koo et al. 2001). Subsequent studies revealed that the GHS-mediated mitigating effects on thymic atrophy are not GH-IGF1 dependent, with its administration failing to raise hormone levels in treated mice. GRL was deemed responsible for stem cell factor (SCF) secretion and the inflammation dampening effects observed in aged mice (Dixit et al. 2007). Its counterpart leptin, was accountable for rising concentrations of IGF1 and KGF. Furthermore, acylated GRL was found to influence thymic rejuvenation beyond a mere increase in cellularity, with a specific TEC-oriented action through the re-establishment of epithelial architecture and the tightening of cortico-medullary boundaries. Concomitantly, significant *in situ* adipocyte loss is seen. The increase in TEC number translated to a higher input of hematopoietic progenitors and augmented T cell output. In conclusion, the potent anti-inflammatory effects of GRL present a major advantage for its clinical use, yet to date, no clinical trials have focussed on its administration for reversal of thymic involution (Taub et al. 2010). Additional studies will be required to further dissect its mode of action.

Reactivation of Resident Stem Cells Using Small RNAs

Although canonical gene expression is critical for instructing differentiation-oriented cell programming, it is certainly not the only process involved. Well known mechanisms of cell regulation include epigenetic, transcriptional, post-transcriptional, translational and post-translational modifications. The mechanism known as RNA interference (RNAi), which acts at the post-transcriptional and translational levels, began with the discovery of what appeared to be RNA-mediated translational inhibition of the *lin-14* mRNA in *C. elegans*, carried out by the small RNA *lin-4* (Lee et al. 1993). Major clarification and deeper insight into this peculiar regulatory mechanism came when Fire and colleagues first reported a dsRNA-mediated silencing of animal genes (Montgomery et al. 1998).

In recent years, miRNAs were shown to be key players in crucial developmental processes such as pluripotency, maintenance and differentiation, including ectoderm, mesoderm, and endoderm derivation (Ivey et al. 2008; Rosa et al. 2009; Tzur et al. 2008; Xu et al. 2009; Eberhart et al. 2008; Labbaye et al. 2008; Muljo et al. 2005; Baltimore et al. 2008; Schickel et al. 2008; Xiao and Rajewsky 2009). Little is known about the regulatory mechanisms that these small RNAs supervise in differentiating cells. Research on the extent to which this network can be artificially exploited to finely tune the expression of known genes implicated in development is still in its infancy. These small molecules have been identified as key ingredients in the developmental recipe to derive terminally differentiated cells from embryonic stem cells (ESCs) (Kanellopoulou et al. 2005).

The broad importance of miRNA regulation within the thymic microenvironment and in thymopoiesis was demonstrated in Dicer-deficient mice, a protein required for the biogenesis of microRNAs (Rupp et al. 2014; Zuklys et al. 2012). In particular, some crucial small RNAs were identified to play a role in thymopoiesis at different stages of T cell differentiation. The ordered expression of several miRNAs were reported to be necessary for the correct development of mature, naïve T cells from hematopoietic precursors: miR-181a and miR-150 were shown to control the DN to DP stage of differentiation, the miR-17-92 cluster was demonstrated to be crucial in the DP to SP thymocyte transition, miR-142 was responsible for Th1 generation, miR-150, miR-155 and miR-146 for their Th2 counterpart, and finally miR-155 and miR-146 for Treg differentiation (Baltimore et al. 2008). An important role in NKT cell development for miRNAs was also demonstrated with a conditional *Dicer* KO in thymocytes, leading to severe NKT cell deficiency due to a block in precursor differentiation (Seo et al. 2010).

Other recent findings have highlighted the importance of thymic miRNA regulation in infection-induced atrophy, using mice challenged with the protozoan *Trypanosoma cruzi* (Linhares-Lacerda et al. 2015). 29 miRNAs were shown to be differentially expressed, with putative involvement in B cell lymphoma 2 (Bcl2) mediated apoptosis, ECM deposition and *Aire* down-regulation. A comparative miRNA expression analysis was conducted by Yaqiong and colleagues using aged, middle aged and young murine thymi (Ye et al. 2014). 42 miRNAs were identified as differentially expressed in the two ageing groups, with up to a third belonging to 11 distinct miRNA clusters. Importantly, both the let-7 and miR-29 families were shown to follow similar expression trends during ageing, while an important link to cancer-related miRNAs that are also involved in senescence was observed.

miR-29a was shown to set the threshold for infection-induced, interferon alpha (IFNα)-mediated thymic involution. This miRNA was found to act on TECs by lowering their sensitivity to IFNα and thereby stopping unnecessary, rapid and permanent loss of thymic cellularity when suboptimal infection signals were present (Papadopoulou et al. 2012). Furthermore, Ucar et al. identified a mutual dependence between *Aire* expression and the presence of miRNAs in the medulla, with a suggested prominent role for miR-29a (Ucar et al. 2013). Additional studies utilising a microprocessor KO in TECs revealed the essential role of DiGeorge syndrome chromosomal region 8 (DGCR8) in terminally differentiated Aire^{+} mTECs, providing solid evidence for their employment of still undefined, canonically-processed miRNAs to establish and maintain a correct negative selection process (Khan et al. 2014).

The difficulty in unravelling the role of small RNAs in thymic biology is certainly tangible. For instance, although miR-205 was found to be highly and uniquely expressed in mTECs both during ontogeny and during adult life, its thymic-associated role remains elusive (Khan et al. 2015). The miR-205 knockout (KO) mouse did not result in any appreciable phenotypic changes in either homeostatic or stress-induced conditions, but this could be due to complementation by other miRNA's. Despite the paucity of knowledge on the involvement of miRNAs in critical processes such as TEC development, homeostasis and ageing, this approach does harbour potential in rejuvenating the thymic remnant as part of future regenerative medicine-based approaches.

Creating de novo *Thymic Tissue*

Adult Stem and Progenitor Cells

Stem and progenitor cell populations are known for their capacity to self-renew and differentiate into multiple lineages and because all tissues are derived from such cells, it has been hypothesised that organs retain endogenous progenitor cells in later life for tissue homeostasis (Slack 2000; Weissman 2000). This is surely the case for epithelial tissues that renew frequently and/or regenerate subsequent to damage (Slack 2000). Due to the difficulties associated with characterising rare and slowly-dividing stem cell populations, many researchers have turned their focus to downstream organ specific progenitors, which are a more abundant population of cells with similar properties. To be able to directly stimulate endogenous stem or progenitor thymic epithelial cells to regenerate the aged or damaged thymus would be a major breakthrough in the field.

While the rationale for utilising adult resident stem cells is strong, and evidence supporting the notion that a pool of such cells exists within many epithelial organs, clearly identification of the phenotype of this often elusive, rare and quiescent type of cell is problematic (Slack 2000). It is well established that embryonic MTS24^{+} TECs are capable of producing a complete thymic microenvironment, but they appear to lose this function in adult life (Bennett et al. 2002; Gill et al. 2002; Rossi et al. 2007a; Wong et al. 2014). The existence of adult TEPC however, is reasonable since the postnatal thymus is capable of regenerating after damage such as chemotherapy at least in young adults and to some extent following castration.

Until recently, the existence of bipotent adult TEC progenitors versus lineage dedicated, unipotent progenitors remained controversial (Chidgey and Boyd 2006). Substantial progress has now been made towards elucidating the bipotent nature of adult TEPC. Wong et al. (2014) identified Foxn1^{+} quiescent, thymic label retaining cells (LRCs) that express high levels of α6 integrin and Sca1, are enriched in stem cell markers, have the capacity to form both mature cortical and medullary epithelium *in vivo* and when cultured *in vitro,* demonstrate colony forming capacity over several passages (Wong et al. 2014). In parallel studies, TEPCs in the adult were identified in a primitive Foxn1^{-} population (Ucar et al. 2014). Alternative markers of epithelial cell subsets are required to further dissect and phenotypically identify these populations more precisely.

Given the key role of Foxn1 in TEC development, alterations in its expression – in particular its overexpression – may aid in regeneration of the thymus in aged patients. In fact, Jin et al. (2014) reactivated Foxn1 expression in native TECs to demonstrate that functional TEPCs persist in the rudiment of the mouse thymus for at least for 6 months (Jin et al. 2014). Furthermore, Bredenkamp et al. (2014a, b) have recently created a conditional *FOXN1* mouse model to test their hypothesis (Bredenkamp et al. 2014a). Using their tamoxifen-inducible *FOXN1* transgenic mouse model, they found that overexpression of *Foxn1* drives thymus regeneration in aged mice, with an overt increase in thymus size seen in 24 month old animals. They found a 2.6-fold increase in total thymocyte numbers following one month of tamoxifen treatment, and demonstrated expansion of all currently defined thymocyte populations. Thymic epithelial architecture was nearly indistinguishable from juvenile controls. Upregulation of Foxn1 in these mice also increased proliferation in total TECs.

Even with some resolution of TEPC identity, significant hurdles remain before such findings can be directly translated to a clinical setting. For instance, these cells display limited reactivation capacity in the adult/aged, and cannot sustain a consistent flow of transit amplifying (TA) cells to generate *de novo* mature TECs. The issue of finding an efficacious way to deliver factors/cytokines/stimulants to re-activate such cells, without eliciting detrimental systemic effects, also cannot be underestimated.

Embryonic and Induced Pluripotent Stem Cells

The use of ESC to generate new TEPCs through *in vitro* directed differentiation, aimed at generating T cells for passive transfer offers an alternative approach to activating resident tissue progenitor cells. Gordon et al. (2004) found ectopic transplantation of isolated mouse pharyngeal endoderm to be sufficient for the creation of a functional thymus, with cTECs and mTECs structurally organised into their respective regions (Gordon et al. 2004). This major finding provided functional evidence on the endodermal origin of TEPCs. Subsequently, several studies have described differentiation protocols to obtain TEPCs from mouse and human cell lines (Parent et al. 2013; Sun et al. 2013; Lai and Jin 2009). This remarkable progress has regrettably been hindered by the relatively low prevalence of target cells generated, amongst various other by-products of differentiation. Furthermore, final differentiation into functional cTEC and mTEC subsets has only been achieved after transplanting cells *in vivo* with other thymic stromal cells. More work is still required in order to dissect the mesendoderm to anterior definitive endoderm to pharyngeal endoderm (ME-ADE-PE) linear progression that eventually leads to the generation of functional, self-renewing thymic progenitors. This in turn would greatly aid in narrowing the focus of future and current differentiation protocols.

The fact that the cells generated from these protocols were able to produce a functional ectopic thymus, sufficient for reconstitution of a vast naïve T cell repertoire upon engrafting in immunodeficient murine recipients, represents a very important milestone. Such grafted mice demonstrated the capacity to reject allogeneic skin grafts, and enhance allogeneic BMT engraftment (Lai et al. 2011;

Parent et al. 2013). More recently, expanding on an established TEPC differentiation protocol via inducing early endoderm progenitors from a knock-in hESC line (Soh et al. 2014), Su and colleagues generated functional thymic epithelial precursors that not only managed to fully differentiate *in vivo*, but also achieve long-term naïve T cell production (Su et al. 2015a). In comparison to the previously published protocols, Su et al. (2015a, b) obtained TEPCs more efficiently through the addition of recombinant Foxn1 and Hoxa3, with T cells produced *in vivo* demonstrating the capacity to mount a physiological response to antigen stimulation. A protective effect was achieved when myelin oligodendrocyte glycoprotein (MOG)-expressing hESC-derived TEPCs were transplanted in experimental autoimmune encephalomyelitis (EAE)-challenged mice, ameliorating EAE pathology through tolerance induction (Su et al. 2015b).

The creation of hESC-derived TEPCs presents some impediment with regards to clinical translation, such as donor cell immunogenicity upon transplantation into the host, and the lingering tumorigenic potential of these pluripotent cells. Applying such differentiation protocols to induced pluripotent stem cells (iPSCs) from patients may overcome some of these obstacles.

iPSCs are reprogrammed somatic cells which demonstrate ESC-like properties (Smith et al. 2009). This reprogramming event more often than not involves the expression of four transcription factors, octamer-binding transcription factor 4 (Oct4), Kruppel-like factor 4 (Klf4), sex determining region Y-box 2 (Sox2) and v-myc avian myelocytomatosis viral oncogene homolog (c-Myc) (OKSM) (Yamanaka 2009). Such technology holds great promise, given that it allows for patient-matched stem cells without the ethical concerns associated with ESCs (Smith et al. 2009). More recently, it was demonstrated that a single transcription factor could induce generation of TEPC, ostensibly eliminating the requirement for a complex series of growth factors to induce differentiation (Bredenkamp et al. 2014a). Forced overexpression of *Foxn1* in mouse embryonic fibroblasts (MEFs) was sufficient to trans-differentiate these cells into functional induced TECs (iTECs), capable of supporting T cell differentiation *in vitro*, and creating a functional thymus graft *in vivo* (Bredenkamp et al. 2014b). Together, these approaches provide a novel paradigm that might prevent, or at least attenuate, the immunological response associated with grafting such cells (Alvarez et al. 2012).

Regeneration via Bioengineered Matrices

Although factor supplementation therapies and stem cell differentiation may have a significant impact on the future of thymic reconstitution, there is an indisputable need for novel, improved methods of administering and delivering such cells with the appropriate supporting molecules. Utilising tissue-engineering to design biologically-compatible scaffolds is but one interesting approach that may support and improve cell survival and differentiation, as well as help deliver and potentiate the effect of important cytokines and growth factors (Wei and Ma 2008).

Supporting Role of Extracellular Matrices in the Thymus

ECM plays a crucial role in the field of regenerative medicine and in establishing proper thymic architecture (Savino et al. 2000). Immunohistological analyses conducted by Berrih et al. (1985) revealed strong expression of type I collagen in the intracapsular and intraseptal fibers of the thymus, whilst expression of type IV collagen, fibronectin and laminin were detected at high levels in the septae, perivascular spaces, and basement membranes (Berrih et al. 1985). Interestingly, cultured TECs demonstrate strong expression of fibronectin, but significantly lower levels of type IV collagen and laminin. These proteins were detected in the intercellular spaces bordering TEC cultures, indicating that ECM elements are synthesised by stromal cells, and possibly mediate intrathymic interactions (Berrih et al. 1985; Savino et al. 1993). Subsequent studies have confirmed the interactions between thymic cells and ECM proteins to be essential for T cell development (Savino et al. 1993, 2000). Thymocyte adhesion to TECs can be enhanced by the presence of fibronectin and laminin, with antibodies against these proteins or their receptors negating the effect (Lannes-Vieira et al. 1993; Lagrota-Cândido et al. 1996; de Mello-Coelho et al. 1997). The role of fibronectin and laminin as adhesion molecules in TEC/TEC and TEC/thymocyte interactions is further implicated by the fact that both cell types express receptors for them (Savino et al. 2000). Interestingly, a recent attempt at recreating a thymocyte differentiation-supportive environment was published, showing initial commitment of HSC toward the T cell lineage (Palamaro et al. 2013). This study utilised a meshwork of skin keratinocytes that were seeded onto a 3D porous scaffold.

Precedent work employing bio-engineering and cell grafting techniques include the utilisation of collagen fibres, hydrogels, and decellularised organ vestiges (Wainwright and Bury 2011; Lickorish et al. 2004; Elvin et al. 2009; Mann et al. 2001). The correct deposition and prevalence of collagen, fibronectin and laminin is most likely essential for TEC function and consequent thymocyte selection (Gameiro et al. 2010). Thus, production of a scaffold that follows these principles could represent a crucial adjuvant in establishing adult TEPC grafts by aiding cell reaggregation.

Biomaterials as Scaffolds for Thymus Regeneration

Biomimetic materials (biomaterials) have the potential to play a fundamental role in regenerative medicine, acting as a base for the development of three-dimensional scaffolds that aid tissue reconstruction (Ma 2008). Two commonly used approaches to scaffold development, which integrate these materials, are protein hydrogels and acellular tissues (Mann et al. 2001; Wainwright 1995). Given the need for adult TEPC reaggregation into 3D thymic microarchitecture structures to produce T cells, it is plausible that such scaffolds may also make a significant contribution to the field of adult thymus reconstruction (van Ewijk et al. 1994).

Hydrogels encompassing matrix proteins or small active peptide domains could be valuable materials for scaffold fabrication, provided that their properties are comparable to ECMs (Mann et al. 2001; Drury and Mooney 2003). Mann et al. (2001) revealed that hydrogels derived from photopolymerisable polyethylene glycol (PEG) can allow for both smooth muscle cell (SMC) migration and growth (Mann et al. 2001). These scaffolds were grafted with peptides KQAGDV and RGDS, in order to promote SMC adhesion. Furthermore, the coupling between a light-based crosslinking technique and fibrinogen hydrogels has given rise to rapidly polymerisable gel sealants and scaffolds that can incorporate cells (Elvin et al. 2009; Sando et al. 2011). These ECM protein-derived gels are biodegradable, and can be used for thymic regenerative purposes since they do not exert short-term cytotoxic effects when made using relevant levels of each crosslinking reagent (Elvin et al. 2009, 2010). More recently, a functional miniature thymus was created using a novel hydrogel system comprising of self-assembling peptide subunits (Tajima et al. 2015). Their EAK16-II/EAKIIH6 hydrogels with anti-EpCAM IgG containing tri-component adaptor complexes tethered TECs, resulting in the formation of 3D TEC clusters. Engraftment of their scaffolds with thymic stroma from 3 to 4 week old mice gave rise to the development of functional T cells 5 weeks post-transplantation into athymic nude mice.

Members of the collagen protein family could also be useful materials for scaffold fabrication, as they demonstrate good biocompatibility and the ability to interact with many cell types (Lickorish et al. 2004; Ramshaw et al. 2009). Lickorish et al. (2004) produced a novel bone graft substitute from this material (Lickorish et al. 2004). They coated their scaffolds with hydroxyapatite, in order to aid cellular attachment. *In vitro* cytotoxicity tests with fibroblasts and rabbit periosteal cells confirmed that their matrices were biocompatible, and allowed for both cell adhesion and proliferation. These findings suggest that it is possible for collagen-based scaffolds to further research on adult thymus regeneration.

Decellularisation of native tissues has likewise been considered as a means of regenerating organs (Buinewicz and Rosen 2004; Ott et al. 2008; Petersen et al. 2010; Uygun et al. 2010). Most notable is the success of AlloDerm, an acellular human dermal tissue matrix that is applicable in a variety of clinical settings (Wainwright 1995; Buinewicz and Rosen 2004). This biomatrix allows for both recellularisation and revascularisation, and, with its low immunogenicity, reveals the potential of decellularisation-based approaches (Wainwright 1995). Several decellularisation techniques have already been established, yet none are flawless (Gilbert et al. 2006).

The potential of decellularised scaffolds in regenerative medicine has previously been demonstrated in a diversity of organs (Ott et al. 2008; Uygun et al. 2010; Buinewicz and Rosen 2004; Petersen et al. 2010). Ott et al. (2008), for instance, have since developed a decellularised whole cadaveric rat heart scaffold, and seeded it with neonatal cardiac cells (Ott et al. 2008). They compared SDS, PEG, and Triton X-100 perfusion treatments of cadaveric rat hearts, finding SDS-treated scaffolds to be most suitable for use based on removal of cell debris. These scaffolds were later perfused with Triton X-100 to wash out trace amounts of SDS. By day 8 post-seeding, their constructs demonstrated pump function equal to 25 % of a 16-week old foetal heart.

Petersen et al. (2010) have also conducted a similar experiment using decellularised adult rat lung instead of heart tissue (Petersen et al. 2010). They perfused CHAPS, sodium chloride, and EDTA into native adult rat lungs to produce their decellularised construct, and recellularised them with neonatal epithelial cells. Following 4 to 8 days in culture, the researchers implanted their seeded constructs into syngeneic rat recipients for short time intervals, where they demonstrated gas exchange through blood gas analysis.

The decellularisation approach has also yielded positive results in the field of thymic regeneration. Fan et al. (2015) have recently decellularised thymi from 3 to 4 week old mice by using freeze-thawing, followed by application of the chemical detergents, sodium dodecyl sulfate (SDS) and Triton X-100 (Fan et al. 2015). Reseeding of these scaffolds via injection with thymic stroma from 2 to 3 week old mice, in conjunction with Lin- progenitors enriched from BM cells, resulted in the production of $CD4^+CD8^+$ DP as well as $CD4^+CD8^-$, or $CD4^-CD8^+$ SP thymocytes *in vitro*. With further optimisation, such scaffolds could allow improvements in the study of T cell development. Engraftment of such scaffolds with thymic stroma also resulted in thymopoiesis by 8 weeks post-engraftment into nude mice. Together, these studies indicate that decellularised scaffolds may further the field of adult thymus regeneration (Fig. 3.2). Their major advantage over collagen-based scaffolds and hydrogels is that their ECM proportions will more closely mimic native thymic tissue.

Possible scenarios for creating self-tolerant T cells could include the co-culture of cord blood, bone marrow or iPS- derived HSCs with OP9-DL4 supporting cells which allows for *in vitro* generation of T cell precursors (Koch et al. 2008), followed by further selection and maturation processes on autologous *de novo* generated TEPC, to induce self-tolerant naïve T cells – as shown in Fig. 3.2a. This T cell pool may then be introduced directly into immunocompromised patients to hasten immune recovery. Alternatively, iPS-derived TEPCs, generated from skin fibroblasts, can be differentiated into mature TECs following *in vivo* transplantation into mesenteric adipose pads or lymph nodes with the aid of a supporting scaffold which would attract BM derived circulating T cell precursors (Fig. 3.2b). These can allow for *de novo* generation of an autologous naïve T cell repertoire.

Conclusion

Regenerative strategies for the thymus are being explored for application in a variety of clinical situations. Recreating a fully functional thymus would not only be beneficial for treating patients with genetic diseases, age-related atrophy, and cancer, but would also be a valuable tool for minimising cancer escape through

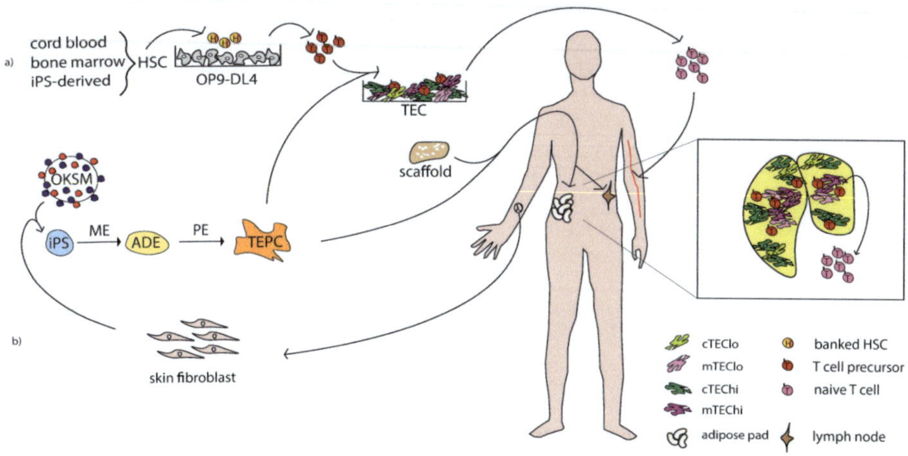

Fig. 3.2 Potential clinical translation of thymic regenerative strategies. (**a**) *In vitro* generation of T cells for passive administration; or (**b**) *in vivo* generation of T cells via transplanted thymic organoids containing *de novo* generated autologous *TEPC*. *ADE* anterior definitive endoderm, *cTEClo* cortical thymic epithelial expressing low MHCII, *cTEChi cTEC* expressing high MHCII, *HSC* hematopoietic stem cell, *iPS* induced pluripotent stem cells, *ME* mesendoderm, *MHCII* major histocompatibility complex class II, *mTEClo* medullary TEC expressing low MHCII, *mTEChi* mTEC expressing high MHCII, *OKSM* Oct4, Klf4, Sox2, c-Myc cassette, *OP9-DL4* OP9 cell line transduced to express delta-like ligand 4, *PE* pharyngeal endoderm, *TEPC* thymic epithelial progenitor cell

increased immune surveillance, boosting central tolerance, the reversal of human immunodeficiency virus (HIV) induced immune deficiency, and the improvement of transplant outcomes by establishing allograft tolerance (Chidgey et al. 2008). Although several regenerative approaches have been trialled, including sex-steroid ablation and growth factor supplementation, problems with patient-to-patient variability in therapeutic response and unwanted side effects remain (Min et al. 2007; Sutherland et al. 2005). New therapeutics and stem cell-based therapies may overcome these issues, and provide a solid more scientifically rational platform to reprogram the immune system to restore thymus functionality. Despite being in its infancy, especially for thymus based regenerative approaches, bioengineering scaffold work may soon be at the forefront of thymic reconstitution therapies. In order to achieve this, a significant amount of work will have to be carried out in order to ensure safety, low toxicity, seeding potential, long-term cell survival, and high *in vivo* deliverability of these devices.

References

Allende ML, Dreier JL, Mandala S, Proia RL (2004) Expression of the sphingosine 1-phosphate receptor, S1P1, on T-cells controls thymic emigration. J Biol Chem 279(15):15396–15401. doi:10.1074/jbc.M314291200

Alpdogan O, Hubbard VM, Smith OM, Patel N, Lu S, Goldberg GL, Gray DH, Feinman J, Kochman AA, Eng JM, Suh D, Muriglan SJ, Boyd RL, van den Brink MR (2006) Keratinocyte growth factor (KGF) is required for postnatal thymic regeneration. Blood 107(6):2453–2460. doi:10.1182/blood-2005-07-2831

Alvarez CV, Garcia-Lavandeira M, Garcia-Rendueles MER, Diaz-Rodriguez E, Garcia-Rendueles AR, Perez-Romero S, Vila TV, Rodrigues JS, Lear PV, Bravo SB (2012) Defining stem cell types: understanding the therapeutic potential of ESCs, ASCs, and iPS cells. J Mol Endocrinol 49(2):89–111

Alves NL, Takahama Y, Ohigashi I, Ribeiro AR, Baik S, Anderson G, Jenkinson WE (2014) Serial progression of cortical and medullary thymic epithelial microenvironments. Eur J Immunol 44 (1):16–22

Andrew D, Aspinall R (2002) Age-associated thymic atrophy is linked to a decline in IL-7 production. Exp Gerontol 37(2–3):455–463. doi:10.1016/s0531-5565(01)00213-3

Annunziato F, Romagnani P, Cosmi L, Lazzeri E, Romagnani S (2001) Chemokines and lymphopoiesis in human thymus. Trends Immunol 22(5):277–281. doi:10.1016/s1471-4906 (01)01889-0

Ara T, Itoi M, Kawabata K, Egawa T, Tokoyoda K, Sugiyama T, Fujii N, Amagai T, Nagasawa T (2003) A role of CXC chemokine ligand 12/stromal cell-derived factor-1/pre-B cell growth stimulating factor and its receptor CXCR4 in fetal and adult T cell development in vivo. J Immunol 170(9):4649–4655

Aspinall R, Andrew D (2000) Thymic involution in aging. J Clin Immunol 20(4):250–256. doi:10.1023/a:1006611518223

Baatar D, Patel K, Taub DD (2011) The effects of ghrelin on inflammation and the immune system. Mol Cell Endocrinol 340(1):44–58

Bachiller D, Klingensmith J, Shneyder N, Tran U, Anderson R, Rossant J, De Robertis EM (2003) The role of chordin/Bmp signals in mammalian pharyngeal development and DiGeorge syndrome. Development 130(15):3567–3578

Baltimore D, Boldin MP, O'Connell RM, Rao DS, Taganov KD (2008) MicroRNAs: new regulators of immune cell development and function. Nat Immunol 9(8):839–845

Baxter RM, Brissette JL (2002) Role of the nude gene in epithelial terminal differentiation. J Invest Dermatol 118(2):303–309

Bennett AR, Farley A, Blair NF, Gordon J, Sharp L, Blackburn CC (2002) Identification and characterization of thymic epithelial progenitor cells. Immunity 16(6):803–814. doi:10.1016/s1074-7613(02)00321-7

Berrih S, Savino W, Cohen S (1985) Extracellular-matrix of the human thymus – immunofluorescence studies on frozen-sections and cultured epithelial-cells. J Histochem Cytochem 33 (7):655–664

Blackburn CC, Manley NR (2004) Developing a new paradigm for thymus organogenesis. Nat Rev Immunol 4(4):278–289. doi:10.1038/nri1331

Blackburn CC, Augustine CL, Li R, Harvey RP, Malin MA, Boyd RL, Miller JF, Morahan G (1996) The nu gene acts cell-autonomously and is required for differentiation of thymic epithelial progenitors. Proc Natl Acad Sci U S A 93(12):5742–5746

Blazar BR, Weisdorf DJ, Defor T, Goldman A, Braun T, Silver S, Ferrara JLM (2006) Phase 1/2 randomized, placebo-control trial of palifermin to prevent graft-versus-host disease (GVHD) after allogeneic hematopoietic stem cell transplantation (HSCT). Blood 108(9):3216–3222

Bleul CC, Boehm T (2005) BMP signaling is required for normal thymus development. J Immunol 175(8):5213–5221

Bleul CC, Corbeaux T, Reuter A, Fisch P, Mönting JS, Boehm T (2006) Formation of a functional thymus initiated by a postnatal epithelial progenitor cell. Nature 441(7096):992–996

Boyd RL, Tucek CL, Godfrey DI, Izon DJ, Wilson TJ, Davidson NJ, Bean AG, Ladyman HM, Ritter MA, Hugo P (1993) The thymic microenvironment. Immunol Today 14(9):445–459

Boyd RL, Soh C-L, Boyd NR, Stanley EG, Chidgey AP (2013) Rewiring immunity: generating a functional thymus from hESCs... are we there yet? Cell Stem Cell 13(2):135–136

Brauchle M, Fässler R, Werner S (1995) Suppression of keratinocyte growth factor expression by glucocorticoids in vitro and during wound healing. J Invest Dermatol 105(4):579–584

Bredenkamp N, Nowell CS, Blackburn CC (2014a) Regeneration of the aged thymus by a single transcription factor. Development 141(8):1627–1637. doi:10.1242/dev.103614

Bredenkamp N, Ulyanchenko S, O'Neill KE, Manley NR, Vaidya HJ, Blackburn CC (2014b) An organized and functional thymus generated from FOXN1-reprogrammed fibroblasts. Nat Cell Biol 16(9):902–908. doi:10.1038/ncb3023

Brennecke P, Reyes A, Pinto S, Rattay K, Nguyen M, Kuchler R, Huber W, Kyewski B, Steinmetz LM (2015) Single-cell transcriptome analysis reveals coordinated ectopic gene-expression patterns in medullary thymic epithelial cells. Nat Immunol 16(9):933–941. doi:10.1038/ni.3246

Brissette JL, Li J, Kamimura J, Lee D, Dotto GP (1996) The product of the mouse nude locus, Whn, regulates the balance between epithelial cell growth and differentiation. Genes Dev 10 (17):2212–2221

Buinewicz B, Rosen B (2004) Acellular cadaveric dermis (AlloDerm): A new alternative for abdominal hernia repair. Ann Plast Surg 52(2):188–194. doi:10.1097/01.sap.0000100895.41198.27

Candi E, Rufini A, Terrinoni A, Giamboi-Miraglia A, Lena AM, Mantovani R, Knight R, Melino G (2007) DeltaNp63 regulates thymic development through enhanced expression of FgfR2 and Jag2. Proc Natl Acad Sci U S A 104(29):11999–12004

Chambers SM, Goodell MA (2007) Hematopoietic stem cell aging: wrinkles in stem cell potential. Stem Cell Rev 3(3):201–211

Chen L, Xiao S, Manley NR (2009) Foxn1 is required to maintain the postnatal thymic microenvironment in a dosage-sensitive manner. Blood 113(3):567–574

Cheng L, Guo J, Sun L, Fu J, Barnes PF, Metzger D, Chambon P, Oshima RG, Amagai T, D-m S (2010) Postnatal tissue-specific disruption of transcription factor FoxN1 triggers acute thymic atrophy. J Biol Chem 285(8):5836–5847

Chidgey AP, Boyd RL (2006) Stemming the tide of thymic aging. Nat Immunol 7(10):1013–1016

Chidgey AP, Layton D, Trounson A, Boyd RL (2008) Tolerance strategies for stem-cell-based therapies. Nature 453(7193):330–337

Chinn IK, Blackburn CC, Manley NR, Sempowski GD (2012) Changes in primary lymphoid organs with aging. Semin Immunol 24(5):309–320. doi:10.1016/j.smim.2012.04.005

Chu Y-W, Memon SA, Sharrow SO, Hakim FT, Eckhaus M, Lucas PJ, Gress RE (2004) Exogenous IL-7 increases recent thymic emigrants in peripheral lymphoid tissue without enhanced thymic function. Blood 104(4):1110–1119. doi:10.1182/blood-2003-10-3635

de Haan G, Van Zant G (1997) Intrinsic and extrinsic control of hemopoietic stem cell numbers: mapping of a stem cell gene. J Exp Med 186(4):529–536

de Mello CV, Villa-Verde DMS, Farias-de-Oliveira DA, de Brito JM, Dardenne M, Savino W (2002) Functional insulin-like growth factor-1/insulin-like growth factor-1 receptor-mediated circuit in human and murine thymic epithelial cells. Neuroendocrinology 75(2):139–150

de Mello-Coelho V, Villa-Verde DMS, Dardenne M, Savino W (1997) Pituitary hormones modulate cell-cell interactions between thymocytes and thymic epithelial cells. J Neuroimmunol 76(1–2):39–49. doi:10.1016/s0165-5728(97)00031-3

Depreter MG, Blair NF, Gaskell TL, Nowell CS, Davern K, Pagliocca A, Stenhouse FH, Farley AM, Fraser A, Vrana J, Robertson K, Morahan G, Tomlinson SR, Blackburn CC (2008) Identification of Plet-1 as a specific marker of early thymic epithelial progenitor cells. Proc Natl Acad Sci U S A 105(3):961–966. doi:10.1073/pnas.0711170105

Derbinski J, Gäbler J, Brors B, Tierling S, Jonnakuty S, Hergenhahn M, Peltonen L, Walter J, Kyewski B (2005) Promiscuous gene expression in thymic epithelial cells is regulated at multiple levels. J Exp Med 202(1):33–45. doi:10.1084/jem.20050471

Dixit VD, Schaffer EM, Pyle RS, Collins GD, Sakthivel SK, Palaniappan R, Lillard JW, Taub DD (2004) Ghrelin inhibits leptin- and activation-induced proinflammatory cytokine expression by human monocytes and T cells. J Clin Invest 114(1):57–66

Dixit VD, Yang H, Sun Y, Weeraratna AT, Youm Y-H, Smith RG, Taub DD (2007) Ghrelin promotes thymopoiesis during aging. J Clin Invest 117(10):2778–2790

Dooley J, Erickson M, Roelink H, Farr AG (2005) Nude thymic rudiment lacking functional foxn1 resembles respiratory epithelium. Dev Dyn 233(4):1605–1612. doi:10.1002/dvdy.20495

Dorshkind K, Horseman ND (2000) The roles of prolactin, growth hormone, insulin-like growth factor-I, and thyroid hormones in lymphocyte development and function: insights from genetic models of hormone and hormone receptor deficiency. Endocr Rev 21(3):292–312

Dorshkind K, Welniak L, Gault RA, Hixon J, Montecino-Rodriguez E, Horseman ND, Gertner JM, Murphy WJ (2003) Effects of housing on the thymic deficiency in dwarf mice and its reversal by growth hormone administration. Clin Immunol 109(2):197–202

Drury JL, Mooney DJ (2003) Hydrogels for tissue engineering: scaffold design variables and applications. Biomaterials 24(24):4337–4351. doi:10.1016/s0142-9612(03)00340-5

Dudakov JA, Goldberg GL, Reiseger JJ, Chidgey AP, Boyd RL (2009a) Withdrawal of sex steroids reverses age- and chemotherapy-related defects in bone marrow lymphopoiesis. J Immunol 182(10):6247–6260

Dudakov JA, Goldberg GL, Reiseger JJ, Vlahos K, Chidgey AP, Boyd RL (2009b) Sex steroid ablation enhances hematopoietic recovery following cytotoxic antineoplastic therapy in aged mice. J Immunol 183(11):7084–7094

Dudakov JA, Khong DMP, Boyd RL, Chidgey AP (2010) Feeding the fire: the role of defective bone marrow function in exacerbating thymic involution. Trends Immunol 31(5):191–198. doi:10.1016/j.it.2010.02.002

Dudakov JA, Hanash AM, Jenq RR, Young LF, Ghosh A, Singer NV, West ML, Smith OM, Holland AM, Tsai JJ, Boyd RL, Van den Brink MRM (2012) Interleukin-22 drives endogenous thymic regeneration in mice. Science 336(6077):91–95

Eberhart JK, He X, Swartz ME, Yan Y-L, Song H, Boling TC, Kunerth AK, Walker MB, Kimmel CB, Postlethwait JH (2008) MicroRNA Mirn140 modulates Pdgf signaling during palatogenesis. Nat Genet 40(3):290–298

Elvin CM, Brownlee AG, Huson MG, Tebb TA, Kim M, Lyons RE, Vuocolo T, Liyou NE, Hughes TC, Ramshaw JAM, Werkmeister JA (2009) The development of photochemically crosslinked native fibrinogen as a rapidly formed and mechanically strong surgical tissue sealant. Biomaterials 30(11):2059–2065. doi:10.1016/j.biomaterials.2008.12.059

Elvin CM, Danon SJ, Brownlee AG, White JF, Hickey M, Liyou NE, Edwards GA, Ramshaw JAM, Werkmeister JA (2010) Evaluation of photo-crosslinked fibrinogen as a rapid and strong tissue adhesive. J Biomed Mater Res Part A 93A(2):687–695. doi:10.1002/jbm.a.32572

Fan Y, Tajima A, Goh SK, Geng XH, Gualtierotti G, Grupillo M, Coppola A, Bertera S, Rudert WA, Banerjee I, Bottino R, Trucco M (2015) Bioengineering thymus organoids to restore thymic function and induce donor-specific immune tolerance to allografts. Mol Ther 23 (7):1262–1277. doi:10.1038/mt.2015.77

Flanagan SP (1966) Nude a new hairless gene with pleiotropic effects in mouse. Genet Res 8 (3):295–309

Gameiro J, Nagib P, Verinaud L (2010) The thymus microenvironment in regulating thymocyte differentiation. Cell Adh Migr 4(3):382–390

Gardiner JR, Jackson AL, Gordon J, Lickert H, Manley NR, Basson MA (2012) Localised inhibition of FGF signalling in the third pharyngeal pouch is required for normal thymus and parathyroid organogenesis. Development 139(18):3456–3466

Gilbert SF (ed) (2010) Developmental biology. Sinauer Associates, Sunderland

Gilbert TW, Sellaro TL, Badylak SF (2006) Decellularization of tissues and organs. Biomaterials 27(19):3675–3683. doi:10.1016/j.biomaterials.2006.02.014

Gill J, Malin M, Hollander GA, Boyd R (2002) Generation of a complete thymic microenvironment by MTS24(+) thymic epithelial cells. Nat Immunol 3(7):635–642. doi:10.1038/ni812

Gnanapavan S, Kola B, Bustin SA, Morris DG, McGee P, Fairclough P, Bhattacharya S, Carpenter R, Grossman AB, Korbonits M (2002) The tissue distribution of the mRNA of ghrelin and subtypes of its receptor, GHS-R, in humans. J Clin Endocrinol Metab 87(6):2988

Godfrey DI, Zlotnik A (1993) Control points in early T-cell development. Immunol Today 14 (11):547–553. doi:10.1016/0167-5699(93)90186-o

Godfrey D, Kennedy J, Suda T, Zlotnik A (1993) A developmental pathway involving four phenotypically and functionally distinct subsets of CD3-CD4-CD8- triple-negative adult mouse thymocytes defined by CD44 and CD25 expression. J Immunol 150(10):4244–4252

Goldberg GL, Sutherland JS, Hammet MV, Milton MK, Heng TSP, Chidgey AP, Boyd RL (2005) Sex steroid ablation enhances lymphoid recovery following autologous hematopoietic stem cell transplantation. Transplantation 80(11):1604–1613

Goldberg GL, Alpdogan O, Muriglan SJ, Hammett MV, Milton MK, Eng JM, Hubbard VM, Kochman A, Willis LM, Greenberg AS, Tjoe KH, Sutherland JS, Chidgey A, Van den Brink MRM, Boyd RL (2007) Enhanced immune reconstitution by sex steroid ablation following allogeneic hemopoietic stem cell transplantation. J Immunol 178(11):7473–7484

Goldberg GL, Dudakov JA, Reiseger JJ, Seach N, Ueno T, Vlahos K, Hammett MV, Young LF, Heng TS, Boyd RL, Chidgey AP (2010) Sex steroid ablation enhances immune reconstitution following cytotoxic antineoplastic therapy in young mice. J Immunol 184(11):6014–6024. doi:10.4049/jimmunol.0802445

Goldberg JD, Zheng J, Castro-Malaspina H, Jakubowski AA, Heller G, van den Brink MRM, Perales M-A (2013) Palifermin is efficacious in recipients of TBI-based but not chemotherapy-based allogeneic hematopoietic stem cell transplants. Bone Marrow Transplant 48(1):99–104

Goldrath AW, Bevan MJ (1999) Selecting and maintaining a diverse T-cell repertoire. Nature 402 (6759):255–262

Gordon J, Wilson VA, Blair NF, Sheridan J, Farley A, Wilson L, Manley NR, Blackburn CC (2004) Functional evidence for a single endodermal origin for the thymic epithelium. Nat Immunol 5(5):546–553

Gordon J, Patel SR, Mishina Y, Manley NR (2010) Evidence for an early role for BMP4 signaling in thymus and parathyroid morphogenesis. Dev Biol 339(1):141–154. doi:10.1016/j.ydbio.2009.12.026

Gray DHD, Ueno T, Chidgey AP, Malin M, Goldberg GL, Takahama Y, Boyd RL (2005) Controlling the thymic microenvironment. Curr Opin Immunol 17(2):137–143

Greenstein BD, Fitzpatrick FT, Adcock IM, Kendall MD, Wheeler MJ (1986) Reappearance of the thymus in old rats after orchidectomy: inhibition of regeneration by testosterone. J Endocrinol 110(3):417–422

Greenstein BD, Fitzpatrick FT, Kendall MD, Wheeler MJ (1987) Regeneration of the thymus in old male rats treated with a stable analogue of LHRH. J Endocrinol 112(3):345–350

Griffith AV, Fallahi M, Venables T, Petrie HT (2012) Persistent degenerative changes in thymic organ function revealed by an inducible model of organ regrowth. Aging Cell 11(1):169–177. doi:10.1111/j.1474-9726.2011.00773.x

Griffith AV, Venables T, Shi J, Farr A, van Remmen H, Szweda L, Fallahi M, Rabinovitch P, Petrie HT (2015) Metabolic Damage and Premature Thymus Aging Caused by Stromal Catalase Deficiency. Cell Rep 12(7):1071–1079. doi:10.1016/j.celrep.2015.07.008

Gruver AL, Hudson LL, Sempowski GD (2007) Immunosenescence of ageing. J Pathol 211 (2):144–156. doi:10.1002/path.2104

Guo J, Feng Y, Barnes P, Huang F-F, Idell S, D-m S, Shams H (2012) Deletion of FoxN1 in the thymic medullary epithelium reduces peripheral T cell responses to infection and mimics changes of aging. PLoS ONE 7(4):e34681

Hattori N (2009) Expression, regulation and biological actions of growth hormone (GH) and ghrelin in the immune system. Growth Horm IGF Res 19(3):187–197

Heng TSP, Goldberg GL, Gray DHD, Sutherland JS, Chidgey AP, Boyd RL (2005) Effects of castration on thymocyte development in two different models of thymic involution. J Immunol 175(5):2982–2993

Heng TSP, Reiseger JJ, Fletcher AL, Leggatt GR, White OJ, Vlahos K, Frazer IH, Turner SJ, Boyd RL (2012) Impact of sex steroid ablation on viral, tumour and vaccine responses in aged mice. PLoS ONE 7(8):e42677. doi:10.1371/journal.pone.0042677

Henson SM, Snelgrove R, Hussell T, Wells DJ, Aspinall R (2005) An IL-7 fusion protein that shows increased thymopoietic ability. J Immunol 175(6):4112–4118

Hirokawa K, Sato K, Makinodan T (1982) Influence of age of thymic grafts on the differentiation of T cells in nude mice. Clin Immunol Immunopathol 24(2):251–262

Hirokawa K, Utsuyama M, Kobayashi S (2001) Hypothalamic control of thymic function. Cell Mol Biol (Noisy-le-grand) 47(1):97–102

Hirokawa K, Utsuyama M, Kikuchi Y (2015) Trade off situation between thymus and growth hormone: age-related decline of growth hormone is a cause of thymic involution but favorable for elongation of lifespan. Biogerontology. doi:10.1007/s10522-015-9590-z

Hollander GA, Wang B, Nichogiannopoulou A, Platenburg PP, van Ewijk W, Burakoff SJ, Gutierrez-Ramos JC, Terhorst C (1995) Developmental control point in induction of thymic cortex regulated by a subpopulation of prothymocytes. Nature 373(6512):350–353. doi:10.1038/373350a0

Holländer G, Gill J, Zuklys S, Iwanami N, Liu C, Takahama Y (2006) Cellular and molecular events during early thymus development. Immunol Rev 209(1):28–46

Holländer GA, Krenger W, Blazar BR (2010) Emerging strategies to boost thymic function. Curr Opin Pharmacol 10(4):443–453

Itoi M, Kawamoto H, Katsura Y, Amagai T (2001) Two distinct steps of immigration of hematopoietic progenitors into the early thymus anlage. Int Immunol 13(9):1203–1211

Itoi M, Tsukamoto N, Amagai T (2007) Expression of Dll4 and CCL25 in Foxn1-negative epithelial cells in the post-natal thymus. Int Immunol 19(2):127–132

Ivey KN, Muth A, Arnold J, King FW, Yeh R-F, Fish JE, Hsiao EC, Schwartz RJ, Conklin BR, Bernstein HS, Srivastava D (2008) MicroRNA regulation of cell lineages in mouse and human embryonic stem cells. Cell Stem Cell 2(3):219–229

Jenkinson EJ, Anderson G, Owen JJ (1992) Studies on T cell maturation on defined thymic stromal cell populations in vitro. J Exp Med 176(3):845–853

Jenkinson WE, Rossi SW, Parnell SM, Agace WW, Takahama Y, Jenkinson EJ, Anderson G (2007) Chemokine receptor expression defines heterogeneity in the earliest thymic migrants. Eur J Immunol 37(8):2090–2096

Jin X, Nowell CS, Ulyanchenko S, Stenhouse FH, Blackburn CC (2014) Long-Term Persistence of Functional Thymic Epithelial Progenitor Cells In Vivo under Conditions of Low FOXN1 Expression. PLoS ONE 9(12):e114842. doi:10.1371/journal.pone.0114842

Johns SA, Soullier S, Rashbass P, Cunliffe VT (2005) Foxn1 is required for tissue assembly and desmosomal cadherin expression in the hair shaft. Dev Dyn 232(4):1062–1068

Kanellopoulou C, Muljo SA, Kung AL, Ganesan S, Drapkin R, Jenuwein T, Livingston DM, Rajewsky K (2005) Dicer-deficient mouse embryonic stem cells are defective in differentiation and centromeric silencing. Genes Dev 19(4):489–501

Kassem M, Marie PJ (2011) Senescence-associated intrinsic mechanisms of osteoblast dysfunctions. Aging Cell 10(2):191–197. doi:10.1111/j.1474-9726.2011.00669.x

Kawakami N, Nishizawa F, Sakane N, Iwao M, Tsujikawa K, Ikawa M, Okabe M, Yamamoto H (1999) Roles of integrins and CD44 on the adhesion and migration of fetal liver cells to the fetal thymus. J Immunol 163(6):3211–3216

Kendall MD, Fitzpatrick FTA, Greenstein BD, Khoylou F, Safieh B, Hamblin A (1990) Reversal of ageing changes in the thymus of rats by chemical or surgical castration. Cell Tissue Res 261 (3):555–564. doi:10.1007/bf00313535

Khan IS, Taniguchi RT, Fasano KJ, Anderson MS, Jeker LT (2014) Canonical microRNAs in thymic epithelial cells promote central tolerance. Eur J Immunol 44(5):1313–1319

Khan IS, Park CY, Mavropoulos A, Shariat N, Pollack JL, Barczak AJ, Erle DJ, McManus MT, Anderson MS, Jeker LT (2015) Identification of MiR-205 As a MicroRNA That Is Highly Expressed in Medullary Thymic Epithelial Cells. PLoS ONE 10(8):e0135440. doi:10.1371/journal.pone.0135440

Ki S, Park D, Selden HJ, Seita J, Chung H, Kim J, Iyer VR, Ehrlich LI (2014) Global transcriptional profiling reveals distinct functions of thymic stromal subsets and age-related changes during thymic involution. Cell Rep 9(1):402–415. doi:10.1016/j.celrep.2014.08.070

Kikuchi T, Ichimiya S, Kojima T, Crisa L, Koshiba S, Tonooka A, Kondo N, Van Der Saag PT, Yokoyama S, Sato N (2004) Expression profiles and functional implications of p53-like transcription factors in thymic epithelial cell subtypes. Int Immunol 16(6):831–841

Koch U, Fiorini E, Benedito R, Besseyrias V, Schuster-Gossler K, Pierres M, Manley NR, Duarte A, MacDonald HR, Radtke F (2008) Delta-like 4 is the essential, nonredundant ligand for Notch1 during thymic T cell lineage commitment. J Exp Med 205(11):2515–2523. doi:10.1084/jem.20080829

Koo GC, Huang C, Camacho R, Trainor C, Blake JT, Sirotina-Meisher A, Schleim KD, Wu TJ, Cheng K, Nargund R, McKissick G (2001) Immune enhancing effect of a growth hormone secretagogue. J Immunol 166(6):4195–4201

Kurobe H, Liu C, Ueno T, Saito F, Ohigashi I, Seach N, Arakaki R, Hayashi Y, Kitagawa T, Lipp M, Boyd RL, Takahama Y (2006) CCR7-dependent cortex-to-medulla migration of positively selected thymocytes is essential for establishing central tolerance. Immunity 24 (2):165–177

Labbaye C, Spinello I, Quaranta MT, Pelosi E, Pasquini L, Petrucci E, Biffoni M, Nuzzolo ER, Billi M, Foà R, Brunetti E, Grignani F, Testa U, Peschle C (2008) A three-step pathway comprising PLZF/miR-146a/CXCR4 controls megakaryopoiesis. Nat Cell Biol 10(7):788–801

Lagrota-Cândido JM, Villa-Verde DMS, Vanderlei JFH, Savino W (1996) Extracellular matrix components of the mouse thymus microenvironment. V. Interferon-[gamma] modulates thymic epithelial cell/thymocyte interactions via extracellular matrix ligands and receptors. Cell Immunol 170(2):235–244. doi:10.1006/cimm.1996.0157

Lai L, Jin J (2009) Generation of thymic epithelial cell progenitors by mouse embryonic stem cells. Stem Cells 27(12):3012–3020

Lai L, Cui C, Jin J, Hao Z, Zheng Q, Ying M, Boyd R, Zhao Y (2011) Mouse embryonic stem cell-derived thymic epithelial cell progenitors enhance T-cell reconstitution after allogeneic bone marrow transplantation. Blood 118(12):3410–3418

Lannes-Vieira J, Chammas R, Villa-Verde DMS, Vannier-dos-Santos MA, Mello-Coelho V, de Souza SJ, Brentani RR, Savino W (1993) Extracellular matrix components of the mouse thymic microenvironment. III. Thymic epithelial cells express the VLA6 complex that is involved in laminin-mediated interactions with thymocytes. Int Immunol 5(11):1421–1430. doi:10.1093/intimm/5.11.1421

Lee RC, Feinbaum RL, Ambros V (1993) The C. elegans heterochronic gene lin-4 encodes small RNAs with antisense complementarity to lin-14. Cell 75(5):843–854

Lickorish D, Ramshaw JAM, Werkmeister JA, Glattauer V, Howlett CR (2004) Collagen-hydroxyapatite composite prepared by biomimetic process. J Biomed Mater Res Part A 68A (1):19–27. doi:10.1002/jbm.a.20031

Linhares-Lacerda L, Palu CC, Ribeiro-Alves M, Paredes BD, Morrot A, Garcia-Silva MR, Cayota A, Savino W (2015) Differential Expression of microRNAs in Thymic Epithelial Cells from Trypanosoma cruzi Acutely Infected Mice: Putative Role in Thymic Atrophy. Front Immunol 6:428. doi:10.3389/fimmu.2015.00428

Linton PJ, Dorshkind K (2004) Age-related changes in lymphocyte development and function. Nat Immunol 5(2):133–139. doi:10.1038/ni1033

Liu C, Ueno T, Kuse S, Saito F, Nitta T, Piali L, Nakano H, Kakiuchi T, Lipp M, Hollander GA, Takahama Y (2005) The role of CCL21 in recruitment of T-precursor cells to fetal thymi. Blood 105(1):31–39. doi:10.1182/blood-2004-04-1369

Liu C, Saito F, Liu Z, Lei Y, Uehara S, Love P, Lipp M, Kondo S, Manley N, Takahama Y (2006) Coordination between CCR7- and CCR9-mediated chemokine signals in prevascular fetal thymus colonization. Blood 108(8):2531–2539

Lynch HE, Goldberg GL, Chidgey A, Van den Brink MRM, Boyd R, Sempowski GD (2009) Thymic involution and immune reconstitution. Trends Immunol 30(7):366–373

Ma PX (2008) Biomimetic materials for tissue engineering. Adv Drug Deliv Rev 60(2):184–198. doi:10.1016/j.addr.2007.08.041

Mackall CL, Fleisher TA, Brown MR, Andrich MP, Chen CC, Feuerstein IM, Magrath IT, Wexler LH, Dimitrov DS, Gress RE (1997) Distinctions between CD8+ and CD4+ T-Cell regenerative pathways result in prolonged T-cell subset imbalance after intensive chemotherapy. Blood 89 (10):3700–3707

Mackall CL, Fry TJ, Bare C, Morgan P, Galbraith A, Gress RE (2001) IL-7 increases both thymic-dependent and thymic-independent T-cell regeneration after bone marrow transplantation. Blood 97(5):1491–1497. doi:10.1182/blood.V97.5.1491

Manley NR, Capecchi MR (1995) The role of Hoxa-3 in mouse thymus and thyroid development. Development 121(7):1989–2003

Manley NR, Capecchi MR (1998) Hox group 3 paralogs regulate the development and migration of the thymus, thyroid, and parathyroid glands. Dev Biol 195(1):1–15

Mann BK, Gobin AS, Tsai AT, Schmedlen RH, West JL (2001) Smooth muscle cell growth in photopolymerized hydrogels with cell adhesive and proteolytically degradable domains: synthetic ECM analogs for tissue engineering. Biomaterials 22(22):3045–3051

Masuda K, Germeraad WTV, Satoh R, Itoi M, Ikawa T, Minato N, Katsura Y, van Ewijk W, Kawamoto H (2009) Notch activation in thymic epithelial cells induces development of thymic microenvironments. Mol Immunol 46(8–9):1756–1767

Melchionda F, Fry TJ, Milliron MJ, McKirdy MA, Tagaya Y, Mackall CL (2005) Adjuvant IL-7 or IL-15 overcomes immunodominance and improves survival of the CD8+ memory cell pool. J Clin Invest 115(5):1177–1187

Meredith M, Zemmour D, Mathis D, Benoist C (2015) Aire controls gene expression in the thymic epithelium with ordered stochasticity. Nat Immunol 16(9):942–949. doi:10.1038/ni.3247

Miller JFAP (2002) The discovery of thymus function and of thymus-derived lymphocytes. Immunol Rev 185:7–14

Min D, Taylor PA, Panoskaltsis-Mortari A, Chung B, Danilenko DM, Farrell C, Lacey DL, Blazar BR, Weinberg KI (2002) Protection from thymic epithelial cell injury by keratinocyte growth factor: a new approach to improve thymic and peripheral T-cell reconstitution after bone marrow transplantation. Blood 99(12):4592–4600

Min D, Panoskaltsis-Mortari A, Kuro-O M, Holländer GA, Blazar BR, Weinberg KI (2007) Sustained thymopoiesis and improvement in functional immunity induced by exogenous KGF administration in murine models of aging. Blood 109(6):2529–2537

Montgomery MK, Xu S, Fire A (1998) RNA as a target of double-stranded RNA-mediated genetic interference in Caenorhabditis elegans. Proc Natl Acad Sci U S A 95(26):15502–15507

Morley JE (2003) Hormones and the aging process. J Am Geriatr Soc 51(7 Suppl):S333–S337

Muljo SA, Ansel KM, Kanellopoulou C, Livingston DM, Rao A, Rajewsky K (2005) Aberrant T cell differentiation in the absence of Dicer. J Exp Med 202(2):261–269

Napolitano LA, Lo JC, Gotway MB, Mulligan K, Barbour JD, Schmidt D, Grant RM, Halvorsen RA, Schambelan M, McCune JM (2002) Increased thymic mass and circulating naive CD4 T cells in HIV-1-infected adults treated with growth hormone. AIDS 16(8):1103–1111

Napolitano LA, Schmidt D, Gotway MB, Ameli N, Filbert EL, Ng MM, Clor JL, Epling L, Sinclair E, Baum PD, Li K, Killian ML, Bacchetti P, McCune JM (2008) Growth hormone enhances thymic function in HIV-1-infected adults. J Clin Invest 118(3):1085–1098

Nehls M, Pfeifer D, Schorpp M, Hedrich H, Boehm T (1994) New member of the winged-helix protein family disrupted in mouse and rat nude mutations. Nature 372(6501):103–107

Nehls M, Kyewski B, Messerle M, Waldschütz R, Schüddekopf K, Smith AJ, Boehm T (1996) Two genetically separable steps in the differentiation of thymic epithelium. Science 272 (5263):886–889

Nowell CS, Farley AM, Blackburn CC (2007a) Thymus organogenesis and development of the thymic stroma. Methods Mol Biol 380:125–162

Nowell CS, Richie E, Manley NR, Blackburn CC (2007b) Thymus and parathyroid organogenesis. In: Lanza R, Langer R, Vacanti J (eds) Principles of tissue engineering, 3rd edn. Elsevier Academic Press, San Diego, pp 647–662

Nowell CS, Bredenkamp N, Tetélin S, Jin X, Tischner C, Vaidya H, Sheridan JM, Stenhouse FH, Heussen R, Smith AJH, Blackburn CC (2011) Foxn1 regulates lineage progression in cortical and medullary thymic epithelial cells but is dispensable for medullary sublineage divergence. PLoS Genet 7(11):e1002348

Okuyama R, Abo T, Seki S, Ohteki T, Sugiura K, Kusumi A, Kumagai K (1992) Estrogen administration activates extrathymic T cell differentiation in the liver. J Exp Med 175 (3):661–669

Olsen NJ, Olson G, Viselli SM, Gu X, Kovacs WJ (2001) Androgen receptors in thymic epithelium modulate thymus size and thymocyte development. Endocrinology 142(3):1278–1283. doi:10.1210/en.142.3.1278

Ortman CL, Dittmar KA, Witte PL, Le PT (2002) Molecular characterization of the mouse involuted thymus: aberrations in expression of transcription regulators in thymocyte and epithelial compartments. Int Immunol 14(7):813–822

Ott HC, Matthiesen TS, Goh S-K, Black LD, Kren SM, Netoff TI, Taylor DA (2008) Perfusion-decellularized matrix: using nature's platform to engineer a bioartificial heart. Nat Med 14 (2):213–221

Palamaro L, Guarino V, Scalia G, Antonini D, De Falco L, Bianchino G, Fusco A, Romano R, Grieco V, Missero C, Del Vecchio L, Ambrosio L, Pignata C (2013) Human skin-derived keratinocytes and fibroblasts co-cultured on 3D poly ε-caprolactone scaffold support in vitro HSC differentiation into T-lineage committed cells. Int Immunol 25(12):703–714

Panoskaltsis-Mortari A, Lacey DL, Vallera DA, Blazar BR (1998) Keratinocyte growth factor administered before conditioning ameliorates graft-versus-host disease after allogeneic bone marrow transplantation in mice. Blood 92(10):3960–3967

Panoskaltsis-Mortari A, Taylor PA, Rubin JS, Uren A, Welniak LA, Murphy WJ, Farrell CL, Lacey DL, Blazar BR (2000) Keratinocyte growth factor facilitates alloengraftment and ameliorates graft-versus-host disease in mice by a mechanism independent of repair of conditioning-induced tissue injury. Blood 96(13):4350–4356

Papadopoulou AS, Dooley J, Linterman MA, Pierson W, Ucar O, Kyewski B, Zuklys S, Holländer GA, Matthys P, Gray DHD, De Strooper B, Liston A (2012) The thymic epithelial microRNA network elevates the threshold for infection-associated thymic involution via miR-29a mediated suppression of the IFN-α receptor. Nat Immunol 13(2):181–187

Parent AV, Russ HA, Khan IS, LaFlam TN, Metzger TC, Anderson MS, Hebrok M (2013) Generation of functional thymic epithelium from human embryonic stem cells that supports host T cell development. Cell Stem Cell 13(2):219–229

Patel SR, Gordon J, Mahbub F, Blackburn CC, Manley NR (2006) Bmp4 and Noggin expression during early thymus and parathyroid organogenesis. Gene Expr Patterns 6(8):794–799

Pennell LM, Galligan CL, Fish EN (2012) Sex affects immunity. J Autoimmun 38(2–3):J282–J291

Perales M-A, Goldberg JD, Yuan J, Koehne G, Lechner L, Papadopoulos EB, Young JW, Jakubowski AA, Zaidi B, Gallardo H, Liu C, Rasalan T, Wolchok JD, Croughs T, Morre M, Devlin SM, Van den Brink MRM (2012) Recombinant human interleukin-7 (CYT107) promotes T-cell recovery after allogeneic stem cell transplantation. Blood 120(24):4882–4891

Petersen TH, Calle EA, Zhao LP, Lee EJ, Gui LQ, Raredon MB, Gavrilov K, Yi T, Zhuang ZW, Breuer C, Herzog E, Niklason LE (2010) Tissue-engineered lungs for in vivo implantation. Science 329(5991):538–541. doi:10.1126/science.1189345

Petrie HT, van Ewijk W (2002) Thymus by numbers. Nat Immunol 3(7):604–605

Petrie HT, Zuniga-Pflucker JC (2007) Zoned out: functional mapping of stromal signaling microenvironments in the thymus. Annu Rev Immunol 25:649–679. doi:10.1146/annurev.immunol.23.021704.115715

Ramshaw J, Peng Y, Glattauer V, Werkmeister J (2009) Collagens as biomaterials. J Mater Sci Mater Med 20:3–8. doi:10.1007/s10856-008-3415-4

Rapaport R, Oleske J, Ahdieh H, Solomon S, Delfaus C, Denny T (1986) Suppression of immune function in growth hormone-deficient children during treatment with human growth hormone. J Pediatr 109(3):434–439

Redelman D, Welniak LA, Taub D, Murphy WJ (2008) Neuroendocrine hormones such as growth hormone and prolactin are integral members of the immunological cytokine network. Cell Immunol 252(1–2):111–121

Rodewald H-R (2008) Thymus organogenesis. Annu Rev Immunol 26(1):355–388

Romano R, Palamaro L, Fusco A, Giardino G, Gallo V, Del Vecchio L, Pignata C (2013) FOXN1: A Master Regulator Gene of Thymic Epithelial Development Program. Front Immunol 4:187

Rosa A, Spagnoli FM, Brivanlou AH (2009) The miR-430/427/302 family controls mesendodermal fate specification via species-specific target selection. Dev Cell 16(4):517–527

Rosenberg SA, Sportès C, Ahmadzadeh M, Fry TJ, Ngo LT, Schwarz SL, Stetler-Stevenson M, Morton KE, Mavroukakis SA, Morre M, Buffet R, Mackall CL, Gress RE (2006) IL-7 administration to humans leads to expansion of CD8+ and CD4+ cells but a relative decrease of CD4+ T-regulatory cells. J Immunother 29(3):313–319

Rossi S, Blazar BR, Farrell CL, Danilenko DM, Lacey DL, Weinberg KI, Krenger W, Holländer GA (2002) Keratinocyte growth factor preserves normal thymopoiesis and thymic microenvironment during experimental graft-versus-host disease. Blood 100(2):682–691

Rossi FM, Corbel SY, Merzaban JS, Carlow DA, Gossens K, Duenas J, So L, Yi L, Ziltener HJ (2005) Recruitment of adult thymic progenitors is regulated by P-selectin and its ligand PSGL-1. Nat Immunol 6(6):626–634. doi:10.1038/ni1203

Rossi SW, Jenkinson WE, Anderson G, Jenkinson EJ (2006) Clonal analysis reveals a common progenitor for thymic cortical and medullary epithelium. Nature 441(7096):988–991. doi:10.1038/nature04813

Rossi SW, Chidgey AP, Parnell SM, Jenkinson WE, Scott HS, Boyd RL, Jenkinson EJ, Anderson G (2007a) Redefining epithelial progenitor potential in the developing thymus. Eur J Immunol 37(9):2411–2418. doi:10.1002/eji.200737275

Rossi SW, Jeker LT, Ueno T, Kuse S, Keller MP, Zuklys S, Gudkov AV, Takahama Y, Krenger W, Blazar BR, Holländer GA (2007b) Keratinocyte growth factor (KGF) enhances postnatal T-cell development via enhancements in proliferation and function of thymic epithelial cells. Blood 109(9):3803–3811

Rupp LJ, Brady BL, Carpenter AC, De Obaldia ME, Bhandoola A, Bosselut R, Muljo SA, Bassing CH (2014) The microRNA biogenesis machinery modulates lineage commitment during alphabeta T cell development. J Immunol 193(8):4032–4042. doi:10.4049/jimmunol.1401359

Sando L, Danon S, Brownlee AG, McCulloch RJ, Ramshaw JAM, Elvin CM, Werkmeister JA (2011) Photochemically crosslinked matrices of gelatin and fibrinogen promote rapid cell proliferation. J Tissue Eng Regen Med 5(5):337–346. doi:10.1002/term.318

Savino W, Villa-Verde DMS, Lannes-Vieira J (1993) Extracellular matrix proteins in intrathymic T-cell migration and differentiation? Immunol Today 14(4):158–161. doi:10.1016/0167-5699(93)90278-s

Savino W, Dalmau SR, Dealmeida VC (2000) Role of extracellular matrix-mediated interactions in thymocyte migration. Dev Immunol 7(2–4):279–291

Savino W, Mendes-da-Cruz DA, Silva JS, Dardenne M, Cotta-de-Almeida V (2002) Intrathymic T-cell migration: a combinatorial interplay of extracellular matrix and chemokines? Trends Immunol 23(6):305–313

Schickel R, Boyerinas B, Park S-M, Peter ME (2008) MicroRNAs: key players in the immune system, differentiation, tumorigenesis and cell death. Oncogene 27(45):5959–5974

Schluns KS, Kieper WC, Jameson SC, Lefrançois L (2000) Interleukin-7 mediates the homeostasis of naïve and memory CD8 T cells in vivo. Nat Immunol 1(5):426–432

Scott IC, Steiglitz BM, Clark TG, Pappano WN, Greenspan DS (2000) Spatiotemporal expression patterns of mammalian chordin during postgastrulation embryogenesis and in postnatal brain. Dev Dyn 217(4):449–456

Seach N, Wong K, Hammett M, Boyd RL, Chidgey AP (2012) Purified enzymes improve isolation and characterization of the adult thymic epithelium. J Immunol Methods 385(1–2):23–34. doi:10.1016/j.jim.2012.07.023

Seo K-H, Zhou L, Meng D, Xu J, Dong Z, Mi Q-S (2010) Loss of microRNAs in thymus perturbs invariant NKT cell development and function. Cell Mol Immunol 7(6):447–453

Slack JMW (2000) Stem cells in epithelial tissues. Science 287(5457):1431–1433. doi:10.1126/science.287.5457.1431

Smaniotto S, de Mello CV, Villa-Verde DMS, Pléau J-M, Postel-Vinay MC, Dardenne M, Savino W (2005) Growth hormone modulates thymocyte development in vivo through a combined action of laminin and CXC chemokine ligand 12. Endocrinology 146(7):3005–3017

Smaniotto S, Mendes-da-Cruz DA, Carvalho-Pinto CE, Araujo LM, Dardenne M, Savino W (2010) Combined role of extracellular matrix and chemokines on peripheral lymphocyte migration in growth hormone transgenic mice. Brain Behav Immun 24(3):451–461

Smith RG, Jiang H, Sun Y (2005) Developments in ghrelin biology and potential clinical relevance. Trends Endocrinol Metab 16(9):436–442

Smith KP, Luong MX, Stein GS (2009) Pluripotency: toward a gold standard for human ES and iPS cells. J Cell Physiol 220(1):21–29. doi:10.1002/jcp.21681

Soh CL, Giudice A, Jenny RA, Elliott DA, Hatzistavrou T, Micallef SJ, Kianizad K, Seach N, Zuniga-Pflucker JC, Chidgey AP, Trounson A, Nilsson SK, Haylock DN, Boyd RL, Elefanty AG, Stanley EG (2014) FOXN1 (GFP/w) reporter hESCs enable identification of integrin-beta4, HLA-DR, and EpCAM as markers of human PSC-derived FOXN1(+) thymic epithelial progenitors. Stem Cell Reports 2(6):925–937. doi:10.1016/j.stemcr.2014.04.009

Sportès C, Hakim FT, Memon SA, Zhang H, Chua KS, Brown MR, Fleisher TA, Krumlauf MC, Babb RR, Chow CK, Fry TJ, Engels J, Buffet R, Morre M, Amato RJ, Venzon DJ, Korngold R, Pecora A, Gress RE, Mackall CL (2008) Administration of rhIL-7 in humans increases in vivo TCR repertoire diversity by preferential expansion of naive T cell subsets. J Exp Med 205 (7):1701–1714

Steinmann GG, Klaus B, Müller-Hermelink HK (1985) The involution of the ageing human thymic epithelium is independent of puberty. Scand J Immunol 22(5):563–575. doi:10.1111/j.1365-3083.1985.tb01916.x

Storek J, Gillespy T, Lu H, Joseph A, Dawson MA, Gough M, Morris J, Hackman RC, Horn PA, Sale GE, Andrews RG, Maloney DG, Kiem H-P (2003) Interleukin-7 improves CD4 T-cell reconstitution after autologous CD34 cell transplantation in monkeys. Blood 101 (10):4209–4218. doi:10.1182/blood-2002-08-2671

Su D-m, Navarre S, W-j O, Condie BG, Manley NR (2003) A domain of Foxn1 required for crosstalk-dependent thymic epithelial cell differentiation. Nat Immunol 4(11):1128–1135

Su M, Hu R, Jin J, Yan Y, Song Y, Sullivan R, Lai L (2015a) Efficient in vitro generation of functional thymic epithelial progenitors from human embryonic stem cells. Sci Rep 5:9882. doi:10.1038/srep09882

Su M, Song Y, He Z, Hu R, Rood D, Lai L (2015b) Administration of embryonic stem cell-derived thymic epithelial progenitors expressing MOG induces antigen-specific tolerance and ameliorates experimental autoimmune encephalomyelitis. J Autoimmun 58:36–47. doi:10.1016/j.jaut.2015.01.002

Sudo K, Ema H, Morita Y, Nakauchi H (2000) Age-associated characteristics of murine hematopoietic stem cells. J Exp Med 192(9):1273–1280

Sultana DA, Tomita S, Hamada M, Iwanaga Y, Kitahama Y, Khang NV, Hirai S, Ohigashi I, Nitta S, Amagai T, Takahashi S, Takahama Y (2009) Gene expression profile of the third pharyngeal pouch reveals role of mesenchymal MafB in embryonic thymus development. Blood 113(13):2976–2987

Sun L, Guo J, Brown R, Amagai T, Zhao Y, D-m S (2010) Declining expression of a single epithelial cell-autonomous gene accelerates age-related thymic involution. Aging Cell 9 (3):347–357

Sun X, Xu J, Lu H, Liu W, Miao Z, Sui X, Liu H, Su L, Du W, He Q, Chen F, Shi Y, Deng H (2013) Directed differentiation of human embryonic stem cells into thymic epithelial progenitor-like cells reconstitutes the thymic microenvironment in vivo. Cell Stem Cell 13(2):230–236

Sutherland JS, Goldberg GL, Hammett MV, Uldrich AP, Berzins SP, Heng TS, Blazar BR, Millar JL, Malin MA, Chidgey AP, Boyd RL (2005) Activation of thymic regeneration in mice and humans following androgen blockade. J Immunol 175(4):2741–2753

Sutherland JS, Spyroglou L, Muirhead JL, Heng TS, Prieto-Hinojosa A, Prince HM, Chidgey AP, Schwarer AP, Boyd RL (2008) Enhanced immune system regeneration in humans following allogeneic or autologous hemopoietic stem cell transplantation by temporary sex steroid blockade. Clin Cancer Res 14(4):1138–1149. doi:10.1158/1078-0432.ccr-07-1784

Tajima A, Liu W, Pradhan I, Bertera S, Bagia C, Trucco M, Meng WS, Fan Y (2015) Bioengineering mini functional thymic units with EAK16-II/EAKIIH6 self-assembling hydrogel. Clin Immunol 160(1):82–89. doi:10.1016/j.clim.2015.03.010

Takahama Y (2006) Journey through the thymus: stromal guides for T-cell development and selection. Nat Rev Immunol 6(2):127–135

Tannenbaum GS, Bowers CY (2001) Interactions of growth hormone secretagogues and growth hormone-releasing hormone/somatostatin. Endocrine 14(1):21–27

Taub DD, Murphy WJ, Longo DL (2010) Rejuvenation of the aging thymus: growth hormone-mediated and ghrelin-mediated signaling pathways. Curr Opin Pharmacol 10(4):408–424

Tuljapurkar SR, McGuire TR, Brusnahan SK, Jackson JD, Garvin KL, Kessinger MA, Lane JT, O'Kane BJ, Sharp JG (2011) Changes in human bone marrow fat content associated with changes in hematopoietic stem cell numbers and cytokine levels with aging. J Anat 219 (5):574–581. doi:10.1111/j.1469-7580.2011.01423.x

Tzur G, Levy A, Meiri E, Barad O, Spector Y, Bentwich Z, Mizrahi L, Katzenellenbogen M, Ben-Shushan E, Reubinoff BE, Galun E (2008) MicroRNA expression patterns and function in endodermal differentiation of human embryonic stem cells. PLoS ONE 3(11):e3726

Ucar O, Tykocinski L-O, Dooley J, Liston A, Kyewski B (2013) An evolutionarily conserved mutual interdependence between Aire and microRNAs in promiscuous gene expression. Eur J Immunol 43(7):1769–1778

Ucar A, Ucar O, Klug P, Matt S, Brunk F, Hofmann TG, Kyewski B (2014) Adult Thymus Contains FoxN1(−) Epithelial Stem Cells that Are Bipotent for Medullary and Cortical Thymic Epithelial Lineages. Immunity 41(2):257–269

Ueno T, Hara K, Willis MS, Malin MA, Höpken UE, Gray DHD, Matsushima K, Lipp M, Springer TA, Boyd RL, Yoshie O, Takahama Y (2002) Role for CCR7 ligands in the emigration of newly generated T lymphocytes from the neonatal thymus. Immunity 16(2):205–218

Unsinger J, McGlynn M, Kasten KR, Hoekzema AS, Watanabe E, Muenzer JT, McDonough JS, Tschoep J, Ferguson TA, McDunn JE, Morre M, Hildeman DA, Caldwell CC, Hotchkiss RS (2010) IL-7 promotes T cell viability, trafficking, and functionality and improves survival in sepsis. J Immunol 184(7):3768–3779

Utsuyama M, Kobayashi S, Hirokawa K (1997) Induction of thymic hyperplasia and suppression of splenic T cells by lesioning of the anterior hypothalamus in aging Wistar rats. J Neuroimmunol 77(2):174–180

Uygun BE, Soto-Gutierrez A, Yagi H, Izamis ML, Guzzardi MA, Shulman C, Milwid J, Kobayashi N, Tilles A, Berthiaume F, Hertl M, Nahmias Y, Yarmush ML, Uygun K (2010) Organ reengineering through development of a transplantable recellularized liver graft using decellularized liver matrix. Nat Med 16(7):814–820. doi:10.1038/nm.2170

van den Brink MRM, Alpdogan O, Boyd RL (2004) Strategies to enhance T-cell reconstitution in immunocompromised patients. Nat Rev Immunol 4(11):856–867

van Ewijk W, Jenkinson EJ, van Soest PL, Owen JJ (1982) Detection of MHC, Thy-1, Lyt-1 and Lyt-2 antigens in the developing mouse thymus. Adv Exp Med Biol 149:241–248

van Ewijk W, Shores EW, Singer A (1994) Crosstalk in the mouse thymus. Immunol Today 15 (5):214–217. doi:10.1016/0167-5699(94)90246-1 van Ewijk W, Wang BP, Hollander G, Kawamoto H, Spanopoulou E, Itoi M, Amagai T, Jiang YF, Germeraad WTV, Chen WF, Katsura Y (1999) Thymic microenvironments, 3-D versus 2-D? Semin Immunol 11(1):57–64

van Ewijk W, Hollander G, Terhorst C, Wang BP (2000) Stepwise development of thymic microenvironments in vivo is regulated by thymocyte subsets. Development 127 (8):1583–1591

Vence LM, Wang C, Pappu H, Anson RE, Patel TA, Miller P, Bassett R, Lizee G, Overwijk WW, Komanduri K, Benjamin C, Alvarado G, Patel SP, Kim K, Papadopoulos NE, Bedikian AY, Homsi J, Hwu WJ, Boyd R, Radvanyi L, Hwu P (2013) Chemical castration of melanoma patients does not increase the frequency of tumor-specific CD4 and CD8 T cells after peptide vaccination. J Immunother 36(4):276–286. doi:10.1097/CJI.0b013e31829419f3

Ventevogel MS, Sempowski GD (2013) Thymic rejuvenation and aging. Curr Opin Immunol 25 (4):516–522

Wainwright DJ (1995) Use of an acellular allograft dermal matrix (AlloDerm) in the management of full-thickness burns. Burns 21(4):243–248. doi:10.1016/0305-4179(95)93866-i

Wainwright DJ, Bury SB (2011) Acellular dermal matrix in the management of the burn patient. Aesthet Surg J 31(7 Suppl):13S–23S

Wei G, Ma PX (2008) Nanostructured Biomaterials for Regeneration. Adv Funct Mater 18 (22):3566–3582

Weissman IL (2000) Translating stem and progenitor cell biology to the clinic: barriers and opportunities. Science 287(5457):1442–1446. doi:10.1126/science.287.5457.1442

Weissman IL, Shizuru JA (2008) The origins of the identification and isolation of hematopoietic stem cells, and their capability to induce donor-specific transplantation tolerance and treat autoimmune diseases. Blood 112(9):3543–3553. doi:10.1182/blood-2008-08-078220

Windmill KF, Lee VW (1999) Influences of surgical castration on the thymus of male rats. J Reprod Immunol 44(1–2):29–39

Windmill KF, Meade BJ, Lee VW (1993) Effect of prepubertal gonadectomy and sex steroid treatment on the growth and lymphocyte populations of the rat thymus. Reprod Fertil Dev 5 (1):73–81

Wong K, Lister NL, Barsanti M, Lim JM, Hammett MV, Khong DM, Siatskas C, Gray DH, Boyd RL, Chidgey AP (2014) Multilineage potential and self-renewal define an epithelial progenitor cell population in the adult thymus. Cell Rep 8(4):1198–1209. doi:10.1016/j.celrep.2014.07.029

Xiao C, Rajewsky K (2009) MicroRNA control in the immune system: basic principles. Cell 136 (1):26–36

Xu N, Papagiannakopoulos T, Pan G, Thomson JA, Kosik KS (2009) MicroRNA-145 regulates OCT4, SOX2, and KLF4 and represses pluripotency in human embryonic stem cells. Cell 137 (4):647–658

Yamanaka S (2009) Elite and stochastic models for induced pluripotent stem cell generation. Nature 460(7251):49–52. doi:10.1038/nature08180

Yano M, Kuroda N, Han H, Meguro-Horike M, Nishikawa Y, Kiyonari H, Maemura K, Yanagawa Y, Obata K, Takahashi S, Ikawa T, Satoh R, Kawamoto H, Mouri Y, Matsumoto M (2008) Aire controls the differentiation program of thymic epithelial cells in the medulla for the establishment of self-tolerance. J Exp Med 205(12):2827–2838

Ye Y, Li D, Ouyang D, Deng L, Zhang Y, Ma Y, Li Y (2014) MicroRNA expression in the aging mouse thymus. Gene 547(2):218–225. doi:10.1016/j.gene.2014.06.039

Zamisch M, Moore-Scott B, D-m S, Lucas PJ, Manley N, Richie ER (2005) Ontogeny and regulation of IL-7-expressing thymic epithelial cells. J Immunol 174(1):60–67

Zlotoff DA, Bhandoola A (2011) Hematopoietic progenitor migration to the adult thymus. Ann N Y Acad Sci 1217:122–138. doi:10.1111/j.1749-6632.2010.05881.x

Zook EC, Krishack PA, Zhang S, Zeleznik-Le NJ, Firulli AB, Witte PL, Le PT (2011) Overexpression of Foxn1 attenuates age-associated thymic involution and prevents the expansion of peripheral CD4 memory T cells. Blood 118(22):5723–5731

Zuklys S, Mayer CE, Zhanybekova S, Stefanski HE, Nusspaumer G, Gill J, Barthlott T, Chappaz S, Nitta T, Dooley J, Nogales-Cadenas R, Takahama Y, Finke D, Liston A, Blazar BR, Pascual-Montano A, Holländer GA (2012) MicroRNAs control the maintenance of thymic epithelia and their competence for T lineage commitment and thymocyte selection. J Immunol 189(8):3894–3904

Chapter 4
Artificial Thymus: Recreating Microenvironmental Cues to Direct T Cell Differentiation and Thymic Regeneration

Mahmood Mohtashami, Shreya Shukla, Peter Zandstra, and Juan Carlos Zúñiga-Pflücker

Abstract The thymus supports the development and differentiation of T cells, which are a central component of the mammalian adaptive immune system. From entry of hematopoietic progenitor cells into the thymus to their complex maturation sequence into naïve T cells, the role of the thymic microenvironment has been the subject of intense study. A pivotal aspect of this process is the activation of Notch receptors on progenitors by Delta-like (Dll) ligands present on thymic epithelial cells. Thus far, two approaches have been taken to create an artificial thymus, or mimic thymic function. One involves an in vitro cell-based system in which several key components are provided, including Dll and cytokines, to induce and support T cell lineage differentiation. The gold standard approach makes use of a bone marrow-derived cell line (OP9), ectopically expressing Dll (OP9-DL). A related method involves an in vitro cell-free system that provides a similar set of required signaling components. The second approach involves the generation of organized thymic tissue, which can be achieved by direct reprogramming of another cell type or via the differentiation of pluripotent stem cells (PSCs), either by the ectopic expression of the thymic master regulatory gene, FoxN1, in fibroblasts or by inducing the differentiation of PSCs using developmental cues, respectively. These approaches share a similar goal, to generate T cells from different sources of stem cells. However, the former takes advantage of cellular or molecular drivers of T-lineage differentiation, while the latter is focused on creating thymic tissues that would support T cell development.

M. Mohtashami (✉)
Sunnybrook Research Institute, Toronto, ON, Canada
e-mail: mahmoodm@sri.utoronto.ca

S. Shukla • P. Zandstra
Institute of Biomaterials & Biomedical Engineering, University of Toronto, Toronto, ON, Canada

J.C. Zúñiga-Pflücker
Sunnybrook Research Institute, Toronto, ON, Canada

Department of Immunology, University of Toronto, Toronto, ON, Canada

T. Watanabe, Y. Takahama (eds.), *Synthetic Immunology*,
DOI 10.1007/978-4-431-56027-2_4

Introduction

The thymus is the primary immune lymphoid organ in vertebrates that provides the proper environment for hematopoietic progenitor cells to differentiate into mature T Lymphocytes (Takahama 2006). This unique capacity to support T cell development is made possible through the function of thymic epithelial cells (TECs) (Anderson and Takahama 2012). TECs are part of the permanent stromal cell structure of the thymus, guiding the transient hematopoietic cells to enter and navigate through the mesh-like structure in close apposition. TECs achieve this through secretion of cytokines, chemokines and expression of cell surface signaling molecules. The interaction between TECs and developing T lymphocytes, called crosstalk, is essential for the normal development of both stromal and hematopoietic cellular components of the thymus (Love and Bhandoola 2011; Lopes et al. 2015). Identifying the nature of signals relayed between these two components is essential for designing an artificial thymus.

Disruption of thymic output of T cells could occur through both physiological and non-physiological processes resulting in low T cell counts in peripheral blood and increased susceptibility to pathogens. The production of T cells in the thymus begins at the embryonic stage in both mice and humans, reaching its peak at adolescence. After this stage, there is a continuous, physiological deterioration of the organ during aging known as thymic involution (Zediak and Bhandoola 2005). Involution results in the decline of T cell production and the reduction of the T cell receptor (TCR) repertoire against foreign antigens, and is thought to lead to an impaired ability to clear pathogens. Thymic involution appears to be influenced by androgens, with its commencement coinciding with the onset of puberty. Androgen-receptor antagonists and orchiectomy in mice both result in an increase in thymic size (Goldberg et al. 2005; Heng et al. 2005). However, because of the transient impact of this treatment (Griffith et al. 2012) and its negative side effects, this is not a permanent solution to thymic decline. Non-physiological events that could damage the thymus are broad. These include the drastic surgical removal of thymus during infant cardiac surgery and effects of diseases that target T cells, such as HIV infection. Even regular viral infections result in a transient atrophy of the thymus (Dooley and Liston 2012). One of the most common forms of non-physiological damage is the effect of radio-chemotherapy for cancer treatment. In older individuals, thymic involution further exacerbates the negative effects of this medical intervention resulting in the decimation of T cells. In addition, radio-chemotherapy also damages the integrity of TECs. In combination, these conditions delay the *de novo* production of T cells during recovery, making the patient highly susceptible to infection, thereby increasing morbidity. While some protection against damage has been reported to occur after treatment with Keratinocyte Growth Factor (KGF, also known as Fibroblast Growth Factor 7) (Kelly et al. 2010) and rejuvenation of TECs by IL-22 in experimental models (Dudakov et al. 2012), these factors remain insufficient for fast recovery with a broad T cell repertoire.

In this chapter, we will discuss two possible solutions to faster and more sustainable T cell recovery. One involves the production of T cells in vitro for adoptive transfer through which lost T cells can be replenished. This strategy would include cell-based systems, such as OP9-DL, that initiate T cell lineage commitment and support the differentiation of hematopoietic stem cells (HSCs). It could also include a cell-free system with the appropriate set of molecules to sustain T lymphopoiesis. The second approach is the regeneration of thymic cell types for transplantation. This could be achieved via ectopic expression of regulatory genes in another cell type or via the directed differentiation of embryonic stem cells (ESC) or induced pluripotent stem cells (iPSCs) to TEC-like cells. The newly generated thymus would replace the involuted, damaged or surgically removed thymus, functioning to attract progenitor cells to enter the organ, supporting their maturation into T cells and ushering them to the periphery (see Fig. 4.1 for a schematic overview).

T Cell Development *in Vitro*

Use of Ex Vivo Tissue Fetal Thymic Organ Cultures (FTOC)

Thymus explants from embryonic day 14–15 mice, also known as fetal thymic organ cultures (FTOC), were the earliest material used for differentiation of HSCs into T lineage cells in culture (Jenkinson et al. 1982). Treatment with purine nucleoside, 2-deoxyguonosine, is selectively toxic to lymphocytes, depleting the T cells, while leaving thymic stromal architecture intact. After washing out the 2-deoxyguanosine, hematopoietic progenitor cells (e.g. derived from day 15 fetal liver or adult bone marrow) from another mouse, possibly with a different genotype, can be introduced into these "empty" FTOCs, which are still able to support T lymphopoiesis (Jenkinson and Owen 1990). This was the first indication that the T cell generation could be replicated in vitro.

The progenitor cells seeding the FTOC go through the normal ontogeny of T cell development as they would in the fully formed thymus. In their early stages of development, the thymocyte progenitors are referred to as double negative cells (DNs) as they still lack both CD4 and CD8 expression. The hallmark of the developing DNs is the sequential acquisition and loss of CD25 and CD44 expression. Many different molecular cues combine at each different substage of DN development to allow progression from DN1 (CD44+ CD25-) to DN2 (CD44+ CD25+), DN3 (CD44- CD25+) and DN4 (CD44- CD25-) before acquiring the expression of both CD8 and CD4 markers at the double positive (DP) stage (Godfrey et al. 1993). Normally, a fully formed thymus contains approximately 3 % DNs, 85 % DPs, 4 % $CD8^+$ single positive (SP) and 8 % $CD4^+$ SPs. Many of the cellular and molecular signals required to advance through T cell development

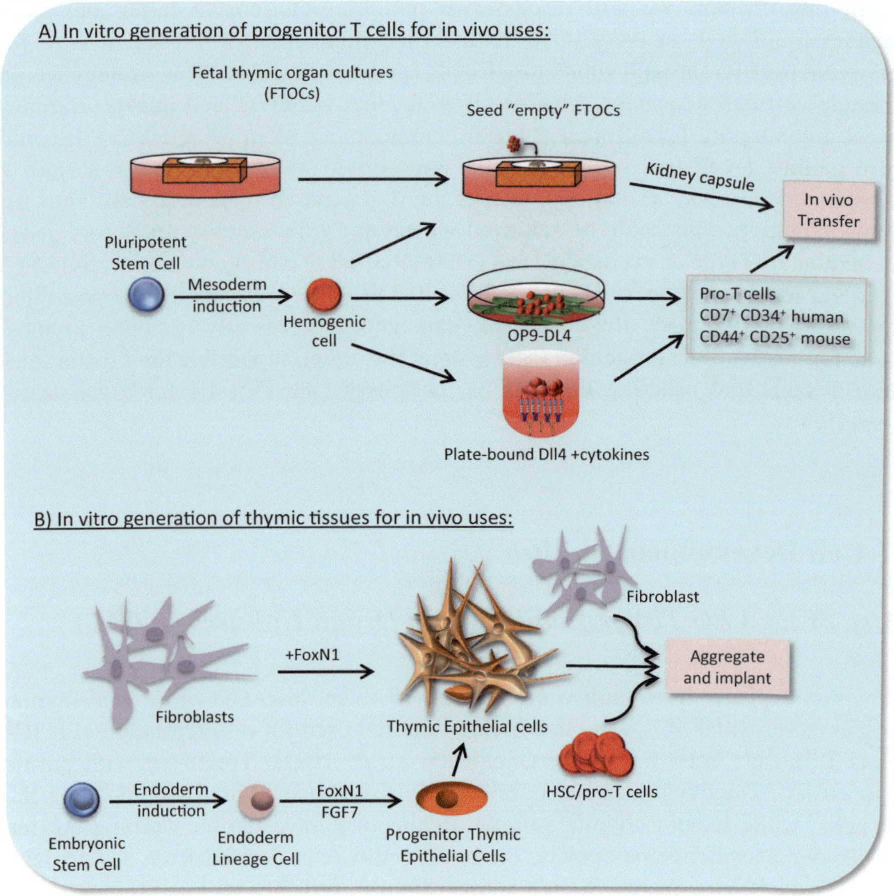

Fig. 4.1 (**a**) Schematic representation of in vitro generation of T cells suitable for adoptive transfer in vivo. Hematopoietic lineage cells derived from pluripotent stem cells or from tissues such as cord blood/fetal liver/bone marrow can be placed with FTOCs or OP9-DL4 cells or in an artificial environment with minimum requirements to give rise to T cells. Reconstituted FTOCs can be implanted under the kidney capsule of the mice. OP9-DL4 cells and plate-bound Dll4 conditions are sufficient to create a minimal T-lineage inducing or thymic-like environment. Pro-T cells derived from these conditions could be transferred into animals for testing or into patients for future therapeutic applications. (**b**) Schematic representation of generation of thymic stromal tissues in vitro. Using genetic manipulation, namely that of FoxN1 overexpression, mouse fibroblasts can be differentiated into induced thymic epithelial cells (iTECs). PSCs can also be manipulated to give rise to thymic tissue, first by their induction to form the endoderm, followed by the addition of FGF7 and other growth factors. Genetic manipulation with the addition of FoxN1 facilitates the differentiation into progenitor thymic epithelial cells. Thymic epithelial cells could then be aggregated with hematopoietic cells (whether HSCs or pro-T cells) along with fibroblasts, yielding the three cellular components of the thymus. The aggregate can then be implanted in vivo and will develop to form a bona fide thymic structures with distinct medullary and cortical regions, and give rise to new T cells in the host

were initially deciphered thanks to the use of FTOCs (Anderson and Jenkinson 2008).

This chimeric system offered an in vitro approach that facilitated many advances for describing the role of TECs versus the hematopoietic cell component during development. Genetic mutations or gene knockouts (KO) cause aberrations in T cell development. In many cases, it is not clear whether the cause of deficiency is attributable to the hematopoietic or/and the stromal component. To verify which component was responsible for the aberration, in vivo adoptive transfer method are typically used. HSCs from the KO or mutant mouse are transplanted into a wild-type (WT) host, irradiated to deplete its own hematopoietic cells, followed by the characterization of the previously described developmental checkpoints. Together with the reciprocal experiment using WT HSC transplants into mutant recipients, the gene of interest's role in thymic or hematopoietic contribution can be determined. Nonetheless, FTOCs provide a faster, more direct and less expensive way to examine the same question; i.e. by transferring WT HSCs into depleted mutant FTOCs and mutant HSCs into WT FTOCs (Anderson and Jenkinson 2008).

In addition, FTOCs provided a more accessible and specific method of studying T cell development. For example it was now possible to study lineage fate choices of a single progenitor cell as it could now be injected into the FTOC (Watanabe et al. 1989; Katsura 2002). Within a few days of injection, the potential of a single progenitor cell to give rise to T cell was determined. In addition, experiments to manipulate or change the thymic milieu, including addition of exogenous reagents such as cytokines, would stay within the confines of the thymic system. In contrast, injection of cytokines systemically in mice to determine its effect on T cell development may cause unwanted side effects that could lead to skewing of the results pertaining to T cell development specifically. The other advantage would be the cost-effectiveness and ease of verifying daily progression of differentiation of T lineage cells when compared to the whole mouse.

Transfer of FTOCs into the subcapsular region of the kidney demonstrated that in vitro-cultured thymic tissue could be transplanted in vivo and function especially to confer tolerance (Jordan et al. 1985). These transplants were able to attract endogenous progenitor cells from the host and generate mature SPs that migrate into the periphery, including the secondary lymphoid organs such as the spleen and lymph nodes. These experiments supported the notion that FTOCs maintained all the essential signaling capacity in order to form a fully functioning thymus. It also demonstrated the great potential for success of transplanting of an in vitro-generated thymus as it can be vascularized, maintained and even developed further.

Despite the capacity of the depleted FTOCs to support T lymphopoiesis, attempts at generating T cells failed when HSCs were cocultured with primary TECs cultured as a monolayer or with cell lines derived from TECs. Instead, in many cases B cells were the resulting cell fate of progenitor cells cultured with TEC monolayers (Montecino-Rodriguez et al. 1996). This led scientists to believe that T lymphopoiesis required a three-dimensional (3-D) environment. This dogma was further strengthened by studies showing the need for 3-D re-aggregation of dissociated FTOCs with HSCs as necessary for the in vitro generation of T cells

(Anderson et al. 1993). In contrast, when dissociated FTOCs were plated on a 2-D culture dish, forming a monolayer, it failed to generate T cells. This suggested that there were one or more key signaling factors that were present in 3-D aggregates, which were absent when cells were cultured in monolayer format.

Requirement of Notch Signaling for T Cell Development

Genetic Studies on mice revealed that Notch signaling was the determining factor in driving T cell fate choice. Deletion of the Notch1 receptor gene, normally expressed in the hematopoietic cells that seed the thymus (as well as HSCs and other early progenitors) leads to a complete absence of T cell development. Instead, B cells appear in the Notch1 conditionally-KO (cKO) thymus (Radtke et al. 1999). Conversely, ectopic expression of the activated form of Notch in hematopoietic cells led to their ability to give rise to T cells within the bone marrow (Pui et al. 1999). The discovery that Notch signaling is unequivocally responsible for modulating T versus B lineage determination led to the conclusion that Notch1 must be activated within the confines of the thymus by one of its ligands. High expression of Notch ligand, Delta-like 4 (Dll4) and to a lesser extent, its ortholog Dll1, in the thymus suggested that they might be responsible for inducing Notch1 activation. While deletion of *Dll4* specifically in TECs led to abrogation of T cell development, *Dll1* deletion with the same specificity had no effect (Hozumi et al. 2004, 2008; Koch et al. 2008). This provided unambiguous evidence that Dll4 is essential in inducing and maintaining the right thymic stromal environment for supporting T cell development.

The pivotal role of Notch/Delta interactions in cell fate determination is not unique to T cell specification. Notch signaling has been shown to be involved in many binary cell fate decisions in different tissues and throughout evolution (Artavanis-Tsakonas and Muskavitch 2010). While humans and mice have four Notch homologs, it is clear that Notch1 is not redundant in embryonic development as deletion of Notch1 gene results in embryonic lethality at E10 (Swiatek et al. 1994). Even in targeted deletion of *Notch1* in hematopoietic cells, expression of Notch2 cannot compensate for the deletion of Notch1 resulting in a complete block in T cell development (Radtke et al. 1999). Dll4 belongs to a family of ligands that include Delta and its homolog Jagged, all of which interact with Notch receptors. In mammals there are 3 members of the delta-like family of proteins and 2 of Jagged (D'Souza et al. 2008). Most members of this family cannot compensate for the deletion or functional deficiency of another. As an examples, deletion of either of the highly homologous Dll1 or Dll4 lead to embryonic lethality, very likely as a result of their non-overlapping expression (Hrabe de Angelis et al. 1997; Duarte et al. 2004; Gale et al. 2004). However, even when expressed in the same pattern of expression in vivo, there is some evidence for functional divergence (Preusse et al. 2015).

A unique feature of the Notch receptor activation is its cleavage after engaging with its ligand, the mechanism of which appears to be common amongst all Notch-Dll/Jagged interactions (Le Borgne et al. 2005; Nichols et al. 2007). Specifically in the thymus during T cell development, Dll4 on TECs interacts with Notch1 on the hematopoietic progenitor cells. Via active endocytosis, Dll4 initiates a pulling action on Notch1 that unmasks its cleavage site close to the plasma membrane. Following two consecutive cleavages within the Notch peptide, the activated Notch intracellular domain (NICD) is released intracellularly in the cytoplasm. NICD then is transported to the nucleus through its nuclear localization signal where it interacts with a DNA binding factor called RBPJ in mice (CBF1 in humans) to initiate the expression of T cell specific genes, including *Deltex*, *preTα*, *HES1*, etc. Failure to engage RBPJ or genetic deletion of *RBPJ* in hematopoietic cells results in a complete block in T cell development (Tanigaki and Honjo 2007).

Hematopoietic Stem/Progenitor Cell (HSPC)-Derived T-Cells Using OP9-DL Cells

The discovery of the Notch receptor's central role in T cell development implied that activation of Notch would have to be a main feature of any artificial thymus capable of deriving T cells from HSCs. It also pointed out that one of the main differences between bone marrow stromal cells and thymic stromal cells maybe the expression of a Notch ligand with the ability to engage and activate Notch1 on hematopoietic cells. This became clearer through the transfer of hematopoietic cells that ectopically expressed Dll4 into wildtype recipient mice. While the cells expressing Dll4 themselves did not differentiate into T cells, they induced the differentiation of endogenous progenitor cells in the bone marrow to adopt a T cell fate through direct contact (Dorsch et al. 2002).

As mentioned above, it was well known that cocultures of HSCs with feeder cells, including those of TEC origin, were unable to promote differentiation of progenitor cells into T cells, but could drive their differentiation to various myeloid, erythroid and even B lymphocyte lineages. A particular murine bone marrow-derived adherent cell line, called OP9, had been widely used for efficient B-lymphocyte generation and proliferation. Other bone marrow-derived cells such as MS5 and S17 were also in use to promote B cell differentiation. However, the great advantage of the OP9 cells is their genotypic background of *op/op*, bearing a functionally defective form of macrophage colony-stimulating factor (M-CSF), a cytokine that strongly promotes myelopoiesis (Nakano et al. 1994). This deficiency makes the OP9 cells remarkably suitable for B cell differentiation, with no competition with myelopoiesis. When OP9 cells were engineered to ectopically express the Notch ligands Dll1 or Dll4 to establish OP9-DL1 and OP9-DL4 cells, respectively, HSC to B lymphocyte differentiation was completely blocked (Mohtashami et al. 2010; Schmitt and Zuniga-Pflucker 2002). Remarkably, instead, T cells were generated. These cultures recapitulated the events that occur within the thymus during T cell commitment at the DN stage, maturation to the DP stage and finally emergence of SP cells. The OP9-DL coculture system's efficiency at T lineage differentiation and proliferation made it an ideal system for experimental manipulation. This could include introduction of transgenes, easy and accessible addition of inhibitors, harvesting large numbers of cells for adoptive transfer and performing limited dilution assays to determine cell fate decisions (Schmitt and Zuniga-Pflucker 2006; Zuniga-Pflucker 2004; de Pooter and Zuniga-Pflucker 2007).

Part of the reason for its effectiveness is that similar to thymic stromal cells, OP9 cells provide key factors that support T lymphopoiesis, including interleukin 7 (IL-7) and stem cell factor (SCF). Genetic deletion of either IL-7 or SCF in mice has negative consequences for both T and B lymphopoiesis, effecting both their survival and proliferation (von Freeden-Jeffry et al. 1995; Rodewald et al. 1995). A double deficiency of IL-7 and SCF receptors blocks T cell development completely (Rodewald et al. 1997), demonstrating the importance of these factors. Flt3-L addition to the coculture allows for better proliferation of the hematopoietic progenitor cells (Wang et al. 2006). These studies present a case for IL-7, SCF and Flt3-L to be included when reconstructing a thymic artificial system. However, as in many biological processes, a right balance must be struck as the addition of exogenous SCF or IL-7 to the coculture or their overexpression in vivo maintains the hematopoietic cells in their progenitor state and actually inhibits their differentiation into lymphocytes (Wang et al. 2006; El Kassar et al. 2004).

While there are remarkably many similarities between the OP9-DL system and the thymus, there are also many differences. The thymus is an organ that is organized in a 3-D web-like structure with an outer cortical and the inner medullary regions, with resident TECs ushering the hematopoietic cells through its intricate niches. In contrast, OP9-DL cells are cultured in a 2-D monolayer format that is uniform with no distinct regions. The migration of hematopoietic cells within the OP9-DL culture is therefore functionally non-consequential because of this uniformity. Despite these differences and the lack of a distinct medullary region where SP cells normally mature, OP9-DL cells allow for the emergence of $CD8^+$ SP T cells with a broad and diverse repertoire of T cell receptors with antigenic specificity, similar to that of thymic $CD8^+$ SPs (Dervovic et al. 2012, 2013). These in vitro-generated $CD8^+$ T cells are functionally equivalent to their thymic counterpart in that they respond to activating signals, and can be negatively deleted at the right concentration of a recognizing peptide. In addition, $CD8^+$ SPs in culture demonstrate tolerance for both self and OP9 cells MHCI. These results indicated that CD8 $^+$ SPs generated in culture are negatively selected in culture against both the stromal cell and the seeding hematopoietic cells, without the need for a medulla (Dervovic et al. 2012; Dervovic et al. 2013).

In contrast to the robust generation of $CD8^+$ SPs, the most significant difference between the thymus and the OP9-DL system is the paucity of $CD4^+$ SP generation. This could be attributable to the absence of MHCII expression by OP9 cells and in hematopoietic cells that would interact with CD4. However, when the transcription factor CIITA was ectopically expressed in OP9-DL cells, inducing the expression of MHCII molecules, no detectible effect on $CD4^+$ SP generation was observed (Dervovic 2011). Therefore it is likely that additional factors, yet to be identified in the thymic medulla (and absent in the OP9-DL), are required for emergence of CD4 $^+$SPs.

OP9-DL1 vs OP9-DL4 Cells

Although Dll4 is the physiological Notch ligand important for T cell development in vivo, the contrast between OP9-DL1 and OP9-DL4 cells appears to be minimal, yielding similar results when promoting T cell differentiation from HSCs. Dll1 and Dll4 have identical structural domains and share about 70 % identity at amino acid level. Yet, they differentially bind Notch1 and Notch2, both of which are expressed in HSCs. While Dll4 has a higher capacity to bind Notch1 compared to Dll1, it has weak interactions with Notch2, whereas Dll1 has the ability to interact with both Notch1 and Notch2 (Besseyrias et al. 2007). Interestingly, the difference between their signaling capacities becomes apparent when Notch1 deficient HSCs are cocultured on OP9-DL cells, differentiating to T-lineage cells on OP9-DL1 but not on OP9-DL4s. Hence, depending on the expression of Notch molecules in the target hematopoietic cells, their interacting partners may differ.

Further differences between Dll1 and Dll4 are observed when they are expressed at lower levels in OP9 cells or levels that are equivalent to that of expressed physiologically by TECs (Mohtashami et al. 2010). Dll expression in the standard OP9-DL cells is about ten-times higher that the physiological levels. These OP9-DL cells show robust T cell production while completely inhibiting B and myeloid induction. At physiological expression levels, Dll4 is more effective than Dll1 at inducing Notch target genes expression and promoting T-lineage commitment, likely because of its stronger interactions with Notch1. More importantly, however, myeloid cells that would normally be absent on the original OP9 cells (without DL), such as dendritic cells, appear in cocultures with physiological levels of Dll4 as observed in the thymus (Mohtashami et al. 2010). These results were somewhat similar to experiments done on FTOCs where a single progenitor cell was injected into an FTOC but clonally either B and T cells never appeared together without the presence of myeloid cells. Myeloid cells could appear by themselves with B or T cells but never B and T without myeloid (Katsura 2002). This confirmed that myeloid cells require Notch stimulation at levels intermediate to those needed for T and B lymphopoiesis, and that myeloid potential is retained after blocking B cell potential (Wada et al. 2008). Therefore, similar to a true morphogen, by adjusting Dll level in OP9 cells, the fate of the cocultured HSCs can be modulated (Schmitt et al. 2004a).

Adjusting the levels of three essential components of OP9-DL system can in effect change the outcome of the cocultured hematopoietic cells. Dll1 or Dll4 levels dictate cell fate choice between T, myeloid and B cells, with increasing Dll expression (and hence Notch activation) skewing the cell fate decisions more towards T lineage (Mohtashami et al. 2010). Increases in SCF concentration leads to higher proliferation and maintenance of progenitor cells, inhibiting differentiation. Similarly, high IL-7 levels result in proliferation of the pro-T cells at DN2 and DN3 stages, partially inhibiting their differentiation into DPs (Balciunaite et al. 2005). This demonstrates the versatility of the OP9-DL coculture system and how it is amenable to various experimental possibilities, including the addition of various cytokines, inhibitors and stimulants.

Adoptive Transfer of Progenitor T Cells

The clinical objective for an artificial thymus would be to replenish T cells in individuals with T cell deficiency. One of the issues expected involves the number of cells available for transplant, which is often a limiting factor. One of the advantages of the OP9-DL coculture system is its ability to support the production of large number of T cells. In addition, new methods have been described whereby the addition of small molecules, such as StemRegenin and UM171, is able to expand the population of HSCs (Fares et al. 2015). The new methods of HSC proliferation together with the potentially limitless expansion of OP9-DL cells, may resolve the paucity of transplantable cells.

An important questions is which T-lineage cells are able to engraft the recipient. Of note, mouse pro-T cells at the DN2 stage of development still have the potential to be chemotactically attracted and home to the thymus. Based on this property, DN2s were injected into lethally irradiated mice along with HSCs. Co-injection of DN2s and HSCs resulted in the appearance of two populations of T cells in the host thymus, one derived from DN2s and the other from HSCs. Interestingly, HSCs first engrafted the bone marrow before seeding the thymus. Even after DN2 progeny had exited the thymus, there was better thymic engraftment by HSC-derived cells, when compared to injection of HSCs alone (Zakrzewski et al. 2006). The maturation and the number of thymocytes generated after myeloablative irradiation was further improved upon by injection of KGF along with the cells. It appears that the increased contribution of HSCs to the thymus was a result of maintaining and improving the thymic environment. It is now clear that pro-T cells that enter the thymus have the ability to modify the architecture of the thymus inducing the cortical and medullary formation, through TEC/hematopoietic interactions. These interactions are likely mediated at least in part by RANK/RANKL signaling (Ohigashi et al. 2011; Rossi et al. 2007). This recovery of the thymus allows for fuller and faster recovery of T cell production post-irradiation.

The clinical implications of a pro-T cell transplant also include the potential for cancer immunotherapy via the introduction of anti-tumor-specific T cells. The success of chimeric antigen receptors (CAR) in preclinical and clinical trials, showed that T cells can successfully reduce tumor load (Zakrzewski et al. 2008). CARs are normally engineered by fusing the tumor-antigen recognizing portion of the IgG to the intracellular domain of CD3. In effect, tumors can be recognized without the MHC context by CAR-bearing T cells and recruit the same activation complex downstream of CD3 that would normally occur through interaction with T cell receptor. Introduced transgenic CARs can be placed under temporal control and expressed only after DN2 progeny mature and exit the thymus into the periphery. This allows the T cells to develop normally within the thymus but express CAR after exit and expand upon encountering the tumor-antigen in order to mediate tumor destruction.

Remarkably, there appears to be no requirement for MHC-matching when it comes to DN2 adoptive transplant into immunodeficient mice. Allogeneic pro-T cells entering the thymus will go through the normal processes of negative and positive selection with the host MHC and their own MHC. In addition, the injected immature DN2s do not yet bear any TCRs to recognize the host MHC as foreign to instigate graft vs host disease. Therefore, having a different MHC profile than the host would be immaterial in this setting as self-reactive T cells be eliminated through negative selection and the surviving T cells would be host-tolerant. The clinical implications of this finding include universal donors for adoptive transfer that can avoid the delays and costs associated with finding a histocompatible donor. Also, it opens up the possibility that hematopoietic component and organs donated from the same individual may now be grafted without the risk of rejection.

Human T Cell Generation in Vitro

Despite the fact that OP9 cells and Dlls expressed by OP9-DL cells are mouse-derived, OP9-DLs were successful in generating *human* T cells from umbilical cord blood (UCB)-derived HSCs. The success of cross-species support demonstrated that mouse Dll1 and Dll4 could interact with human NOTCH, and there are likely other factors expressed by OP9 cells that function across the species barrier. As a result, similar to the mouse model, OP9-DL could also recapitulate the human T cell development and ontogeny. Shortly after UCB-derived HSCs ($CD34^{+}CD38^{lo/-}$) are placed on OP9-DL, they acquire CD7 expression, a first sign of T-lineage differentiation in human hematopoietic cells and retained throughout T cell ontogeny. This is followed by the loss of CD34 expression and gain of CD5 and subsequently CD1a, corresponding to commitment to the T lineage. Similar to the mouse model, human T cells reach a $CD4^{+}CD8^{+}$ DP stage as they differentiate further, but the length of time needed reach this stage in vitro is about 20 days, twice as long as mouse fetal liver-derived HSCs require to reach this stage (Van Coppernolle et al. 2009; La Motte-Mohs et al. 2005). Further differentiation on OP9-DL cells, yields mature SP cells, and similar to the mouse system with a strong bias towards the $CD8^{+}$ SP fate. In vitro-generated $CD8^{+}$ SPs, the majority of which are $CD3^{+}$ $TCR\alpha\beta^{+}$, show similar levels of CD8-specific transcription factors when compared to ex-vivo human UCB-derived T cells (Awong et al. 2011). When activated through CD3/CD28 stimulation, similar to CD8 cytotoxic cells, the in vitro-derived CD8 cells acquire the expression of the cytotoxic agent Granzyme B and release the effector cytokine, Interferon-γ. These results along with the expression of specific cell surface marker of effector T cell activation, pointed to the functionality of the in vitro-generated human $CD8^+$ SP cells.

There are further parallels between the properties of mouse and human T cells, including their rate of differentiation. Both mice and human progenitors from a younger source (fetal liver in mice, UCB in human) possess a higher capacity for in vitro T cell development than adult bone marrow-derived HSCs. In addition, these progenitors also generate greater numbers of T lineage cells, reaching developmental milestones faster than bone marrow-derived HSCs. In terms of stage of development, murine pro-T cells defined as DN2 ($CD44^+CD25^+$) are equivalent to human T-lineage cells expressing $CD34^+CD7^+$. Similar to their mouse counterpart, these human pro-T cells can seed the thymus after being injected into immunodeficient NOD-SCID-gamma(c)null (NSG) mice (Awong et al. 2009; Ito et al. 2002). Once in the thymus, these human pro-T cells undergo a normal program for T cell differentiation. Surprisingly, through their interaction with the mouse TECs, the pro-T cells initiate the restructuring of the NSG mouse thymic architecture. Before injection, NSG mice had a disorganized, underdeveloped thymus, reminiscent of d15 fetal thymus before the appearance of cortical and medullary regions. The emergence of distinct medullary and cortical regions in the adult thymus of NSG mice, after pro-T cell entry, signified the maturation of the thymus as a result of cross-talk. This is most likely responsible for facilitating the faster entry of bone marrow-derived progenitors to home and enter the thymus as the earlier wave of pro-T cells mature and exit the thymus (Awong et al. 2013).

Although UCB-derived HSCs proved very amenable to OP9-DL growth and T cell differentiation, it appears that the nature of the hematopoietic cells and their source have a great influence on the success of T cell differentiation. OP9 cells and OP9-DL provides a suitable microenvironment for mouse ESCs to become hematopoietic and then convert to T-lineage cells (Schmitt et al. 2004b; Holmes and Zuniga-Pflucker 2009). Despite bearing the right markers of $CD45^+CD34^+$, these cells derived from human ESCs do not readily develop into T cells when placed on OP9-DL cells, while successfully differentiating into other hematopoietic lineages (Vodyanik et al. 2006). Proper human ESC-derived hematopoietic progenitors that have the potential to differentiate into T cell were only generated when an ESC-derived structure referred to as the hematopoietic zone was first placed on OP9 cells (without Dll). These hematopoietic zones can only be identified morphologically (Timmermans et al. 2009). Additionally, another approach made use of defined culture conditions with embryoid body formation from hESCs to generate $CD34^+CD45^-$ cells that could be induced to adopt the T-lineage on OP9-DL4 cells (Kennedy et al. 2012). This suggests that cell types other that OP9s, preferably human cell lines, may be more suitable for generating hemogenic cells from hESCs that can be differentiated into T cells.

Use of Alternative Stromal Cells to Generate T Cells in Vitro

Other than in OP9 cells, ectopic expression of Dll1 or Dll4 in many other stromal cell lines, such as S17, MS5, Tst-4 and PA6 has been effective in supporting T cell development (Abe et al. 2010; Armstrong et al. 2009; Ikawa et al. 2010; Jaleco et al. 2001). However, none have proven to be as efficient as OP9-DL. There are limitations to the OP9-DL coculture system including the fact that OP9 cells are mouse-derived and not yet approved for clinical use. In addition, mature human $CD8^+$ SPs, generated using the OP9-DL system, may show reactivity against the host upon adoptive transfer. As demonstrated in mouse HSC/OP9-DL cocultures, $CD8^+$ SPs generated are non-reactive against the OP9 stromal cell MHC and self-MHC through the process of negative selection (Dervovic et al. 2012, 2013). To translate this into clinical setting, and to avoid graft-vs-host reactions, either the hematopoietic cells or the stromal layer must be immunocompatible with the genetic background of the recipient host. To satisfy these shortcomings, an alternative human stromal cell system that would be patient-specific may be required for clinical applications.

One way of achieving this would be to ectopically express Dll4 in the recipient patient's primary cells, such as the easily accessible dermal fibroblasts, instead of the more intrusive attempt to extract bone marrow stromal cells. Surprisingly, expression of Dll4 in mouse TEC monolayers and primary dermal fibroblasts (mFibro-DL4) conferred the ability to support the generation of T-lineage cells (Mohtashami et al. 2013; Mohtashami and Zuniga-Pflucker 2006; Smeets et al. 2013). While not as efficient as the bone marrow-derived OP9-DL, DN2 cells generated in mFibro-DL4/HSC cocultures did nevertheless successfully engraft into immunodeficient mice. The injected cells homed to the thymus and continued to differentiate into mature T cells, which then populated the secondary lymphoid organs such as the spleen (Mohtashami et al. 2013; Smeets et al. 2013). Additionally, mFibro-DL4/HSC coculture could also give rise to $CD8^+$ SP cells.

The success of the mFibro-DL established that stromal cells other than the ones derived from the bone marrow, and specifically dermal fibroblasts could be used to support T cell development in vitro. It also proved that an autologous T cell coculture system could be designed where both the supporting stromal cells and the T cells are derived from the same individual or genetic background. While so far, human fibroblasts have proven difficult to manipulate into creating such system (Mohtashami et al. 2013), there is the potential to generate human pro T cells and $CD8^+$ SPs in culture for use in transplantation.

Cell-Free Support System DL-Fc

Engineering robust technologies to generate progenitor T cells from allogeneic HSCs and PSCs could provide a scalable source of transplantable immune cells and offer important tools for modeling immune diseases. As the OP9-DL system is currently not readily clinically translatable, one approach can be to move to a stromal cell-free system where stem cells are exposed to the minimum molecular requirements for T cell development.

Apart from the cytokines Flt3L, IL-7 and SCF, Dll proteins are indispensable to activate the Notch receptor on HSCs and drive T-lineage commitment and differentiation. As Dll is a transmembrane protein, its expression in culture and purification of the whole protein were shown to be problematic. Instead, the extracellular domain of Delta-like proteins have been engineered to be fused to either a series of myc epitopes (Delta-ext-myc) or the heavy chain domains 2 and 3 of IgG, which are the segments that constitute the Fc domain of IgG (DL-Fc). Delta-ext-myc and DL-Fc are monomeric and dimeric forms of the Delta-like proteins, respectively. DL-Fc is dimeric as the Fc domains are cross-linked to each other by disulphide bonds, forming a dimer when expressed. In both cases, immobilization of the Delta fusion proteins to a surface is key for induction of Notch signaling. Of note, engineered Delta fusion proteins are inhibitory to Notch signaling when not immobilized and added in solution (Varnum-Finney et al. 2000). The DL-Fc ligands are either adsorbed directly to the plate surface, or captured via anti-Fc antibodies. When culturing human $CD34^+$ cells, the fibronectin fragment CH-296 can be coated along with DL1-Fc to promote stronger intracellular signals (Delaney et al. 2010).

Delta fusion proteins have been successfully used to understand the molecular mechanisms that drive progenitor T cell commitment in mouse models of blood development. Ikawa *et al.* used DL4-Fc immobilized on plastic plates via anti-human Ig-Fc antibody to demonstrate that high IL-7 concentrations arrest differentiating thymocytes in the self-renewing DN2mt (myeloid-T) stage via the suppression of the Bcl11b transcription factor (Ikawa et al. 2010). In this feeder-free culture system, they were able to differentiate $Lin^-Sca1^+cKit^+$ HSCs past the TCRβ-selection checkpoint and produce $\alpha\beta TCR^+CD4^+CD8^+$ DP T cells by lowering the concentration of exogenous IL-7 from 10 to 1 ng/mL after one week, demonstrating the importance of providing the proper stimuli and factors at different stages of T cell development (Ikawa et al. 2010).

In addition to high rates of proliferation, murine progenitor T cells generated on engineered Delta fusion proteins are also capable of significant self-renewal. Gehre *et al.* demonstrated that murine DN2-3 T cells generated on DL4-Fc coated plates with no stromal cell support could self-renew for up to 6 months with a doubling time of ~30 h and reconstitute all T cell compartments in vivo in sublethally irradiated $preT\alpha^{-/-}$ mice, including single-positive (SP) CD4, CD8 and $Foxp3^+$ Tregs (Gehre et al. 2015). In addition, they showed that the adoptive transfer of in vitro derived DN2-3 T cells accelerated T cell reconstitution after allogeneic HSC transplantation of all T cell compartments similar to injection of human progenitor T cells with HSCs (Awong et al. 2013).

The field had made significant progress in understanding human T cell development and expanding UCB-derived hematopoietic stem/progenitor cells (HSPCs) for transplantation using DL-Fc fusion proteins. Delaney *et al.* were the first to use engineered recombinant Notch ligand DL1-Fc with defined serum-free media supplementation in a Phase I clinical trial to increase the number of cord blood progenitor cells capable of rapid myeloid engraftment as early as 3 weeks after transplantation (Delaney et al. 2010). They had previously shown that the expansion and in vivo marrow repopulating ability of enriched $CD34^{+}CD38^{-}$ UCB-derived HSPCs can be enhanced in a dose-dependent fashion using DL1-Fc (Delaney et al. 2005). One of the ways DL1-Fc regulates HSPC fate is by reducing membrane-bound IL-6 receptor on the surface of $CD34^{+}$ progenitor cells in a dose dependent manner (Csaszar et al. 2014). This Notch-mediated cell extrinsic regulation of HSPCs prevents them from being IL-6 responsive and thereby alters intercellular feedback networks to reduce skewing towards the myeloid cell fate.

Interestingly, DL1-Fc expanded UCB grafts did not contribute to $CD3^{+}$ T cell engraftment in any of the Phase I clinical trial subjects (Delaney et al. 2010). As DL1-dependent expansion of $CD34^{+}$ cells is mediated by the Notch2 receptor, T cell development was not affected, as it is dependent on Notch1 signaling regulated in the thymus primarily by the DL4 ligand. To address this, the Andre-Schmutz group has been developing an ex vivo culture system using immobilized DL4-Fc to produce early T cell progenitors from UCB-derived $CD34^{+}$ cells for accelerating T cell reconstitution after UCB transplantation (Reimann et al. 2012). They produced a mixture of early thymic progenitors ($CD34^{+}CD45RA^{+}CD7^{+}$), proT1 ($CD7^{++}$ $CD5^{-}$), proT2 ($CD7^{++}CD5^{+}$) and preT ($CD7^{++}CD5^{+}CD1a^{+}$) subsets after 2 weeks of culture on immobilized DL4-Fc using media containing 20 % fetal bovine serum. They found that co-transplantation of a mixture of these DL4-Fc expanded cells with an un-manipulated $CD34^{+}$ UCB graft accelerated peripheral T cell reconstitution in the NSG mouse model compared to transplanting either cell type alone (Reimann et al. 2012). Potential intercellular interactions between the early engrafting pro-T subsets and the post-irradiation injured thymic niche may help improve thymic architecture for subsequent $CD34^{+}$ progenitor cell derived-engraftment (Awong et al. 2013). Thus, pro-T cell expansion mediated through DL4-Fc fusion protein promotes T cell reconstitution and accelerates *de novo* T cell development in vivo.

DL4-Fc protein shows the highest avidity for Notch1 receptor-Delta fusion protein interactions on thymocytes (Besseyrias et al. 2007; Heinzel et al. 2007). In particular, DL4-Fc fusion proteins can be used to stain for Notch receptor expression in a concentration and calcium-dependent manner on primary thymocytes (Heinzel et al. 2007). As discussed earlier, it is possible to titrate the amount of Dll expressed on OP9 stromal cells by sorting for different levels of transduced Dll1 or Dll4 construct expression (Mohtashami et al. 2010). However, it is still difficult to quantify the exact amount of ligand present at steady-state on the surface of TECs due to the dynamic nature of ligand endocytosis-mediated recycling (Shah et al. 2012; Shah and Zuniga-Pflucker 2012). For an engineered application, Delta fusion proteins may provide a more reproducible and scalable choice for Notch activation (Dallas et al. 2005).

Delta fusion proteins can also be clustered into higher molecular complexes that could then signal in a soluble form (Heinzel et al. 2007). Bleul and colleagues showed functionality of DL1-Fc pre-clustered with cross-linking biotinylated anti-

Fc antibodies and streptavidin when added to thymic multipotent precursors (TMPs or $Lin^{-}CD25^{-}CD117^{high}CCR9^{EGFPhigh}$) cultured on OP9 stromal cells. While this strategy was not completely stromal cell-free, TMPs were able to generate DN3 thymocytes with increasing concentrations of pre-clustered DL1-Fc via a γ-secretase dependent mechanism. The ability to form complex soluble Notch ligands makes it possible to titrate the Notch signal strength and directly compare multiple Notch ligands to each other. This is not possible with a stromal-based Notch presentation system as it is difficult to enumerate functional Notch ligands on stromal cells, the mRNA levels of different Notch ligands are hard to control simultaneously in the same stromal system and the overexpression of one Notch ligand may affect the expression level of other Notch ligands (Lehar et al. 2005).

However, there are drawbacks to the soluble Notch ligand complexes, as this approach appears not to work with multipotent precursors from the bone marrow that require stronger Notch signals to initiate T cell commitment. Even with TMPs there were NK cells produced and thus these soluble complexes could not completely suppress non-T lineages. Thus, for current in vitro applications, immobilized Delta fusion proteins are the best choice for generating progenitor T cells in a scalable manner from HSCs and PSCs.

Artificial Thymus Development

FoxN1 and Nude Mice: A Model for Discovery

In vitro studies discussed so far have been able to exploit the molecular mechanisms necessary for T cell development. However, at this point in time, full T cell recovery in immunodeficiency models is still ultimately dependent on the presence of a functional thymus. For adoptive therapy, injected pro-T cells would have to migrate to the host thymus by chemo-attraction and then finalize differentiation and maturation within the thymus. Adoptive transfer would be unsuccessful in cases where the patient's thymus has already been removed, for example because of cardiac surgery, or rendered non-functional due to genetic deficiencies or disease. More commonly, if the thymus has undergone age-related involution and subjected to radio-chemotherapy, it may no longer efficiently attract progenitors and support T cell development. For such cases, an alternative approach to generate and propagate mature T cells is required, involving the regeneration and/or transplantation of the whole thymus.

At the center of attempts to regenerate a thymus has been the transcription factor, FoxN1, a Forkhead-box (F-box) gene family member. The F-box gene family is comprised of a diverse group of "winged-helix" transcription factors and is implicated in various developmental processes, including different aspects of immune regulation (Coffer and Burgering 2004). F-box transcription factors function in directing tissue specific cell fate decisions and this affects pattern formation within whole organs. FoxN1 is no exception to that rule, and, in particular, has a direct and profound effect on thymic development. It is expressed exclusively in TECs and skin epithelial cells. Mutations in *FoxN1* (called *winged helix nude, whn,* at the time of its discovery) result in the hairless or *nude* phenotype in both humans and mice (Blackburn et al. 1996; Frank et al. 1999; Nehls et al. 1994). In *nude* mice, in place of a thymus with differentiated cortical and medullary regions, an undifferentiated anlage is formed that does not express Dll4 or the chemoattractant CCL25 (Calderon and Boehm 2012; Tsukamoto et al. 2005). As a result, these mice develop a severe combined immunodeficiency (SCID) associated with T cell deficiency, due to defect in TECs.

In the absence of thymus-derived T cells in *nude* mice, there is a lack of effective cell-mediated immunity. This property, known as the SCID phenotype, makes *nude* mice particularly suitable for experimental procedures such as xenogenic engraftment, for example in studying human tumor development (Ebos et al. 2009). One could validate the ability of clonogenic T cells to recognize and destroy the target tumor or in combination with various drugs. More importantly, from the perspective of developmental biologists, the thymic rudiment present in *nude* mice provides a clean slate for re-expression of various proteins to determine which in particular are crucial for thymus formation in vivo.

Boehm and colleagues approached the re-formation of the thymus in the *nude* mice from two angles. One was to overexpress the closely related FoxN1 paralog, FoxN4 and determine if it could rescue the SCID phenotype in *nude* mice. This could also determine the role and importance of FoxN1 in thymus formation (Swann et al. 2014). The second approach was to express genes downstream of FoxN1, only in the thymic anlage, to determine which combination was sufficient to reconstruct the thymic function (Calderon and Boehm 2012). As expected, and demonstrated by several different labs, FoxN1 expression in the primordial thymus of *nude* mice gave rise primarily to T cells. In contrast, expression of FoxN4 resulted in generation of both T and B cells, though segregated into different microenvironments within the thymus. B cell generation occurred even when FoxN1 and FoxN4 were co-expressed (Swann et al. 2014). This recapitulated what is observed in the fish, where a primary lymphoid organ generates both B and T cells. This comparison suggested that the F-box transcription factors FoxN1 and FoxN4 control the expression of a different set of downstream factors that contribute to the generation of B or T cells. In effect, one could re-engineer unique types of organs in vivo that could support different hematopoietic cells depending on the regulatory transcription factor expressed. From an evolutionary perspective, it is intriguing that FoxN4 retains its ability to superimpose a different type of cellular outcome in a primary lymphoid organ. However, it does not rule out contributions of the context of the tissue in which each factor is expressed and opens the question of whether they could have the same effect in another tissue. Clearly, FoxN1 expression alone does not establish a thymus when expressed in any cell type, as it is in the skin epithelia, while FoxN4 in mammals is involved in heart development (Cohen and Morrisey 2008).

The flexibility of the *nude* thymic anlage in accommodating the generation of different lymphocyte lineages was shown in another way. In this approach, four factors were expressed singularly in the thymic anlage or in different combinations. These four include, in the order of importance in generating T cells, Dll4, SCF, and the chemoattractants CXCL12 and CCL25 (Calderon and Boehm 2012). Expression of SCF, CXCL12 and CCL25 resulted in development of primarily of B cells with some myeloid cells, while addition of Dll4 to this combination promoted the generation of T cells at the expense of the other lineages. The development of various lineages in this in vivo system paralleled the in vitro OP9-DL/HSC cocultures, in which different levels of Dll expression allowed the differentiation of different sets of hematopoietic lineages. While CCL25 and CXCL12 are not entirely required in vitro, as there is no requirement for chemoattraction in a uniform culture, Dll4 and SCF are the same factors that are added to OP9 cells to support human T cell development. Hematopoietic cells entering the thymus in this system have a positive influence on the expansion of the *nude* thymus. However, the thymocyte/TEC cross-talk was not sufficient to stimulate the formation of a medullary region within the *nude* thymus (Calderon and Boehm 2012). This indicated that while FoxN1 was required for formation of a *bona fide* thymus and its super structure, its function was not limited to regulating the expression of Dll4, CXCL12 and CCL25.

Thymus Regeneration: FoxN1 Master Regulatory Transcription Factor for Reprogramming

As mentioned earlier, many lines of evidence suggest that thymus regeneration is possible. Firstly, the discovery that the single gene, FoxN1, was responsible for the athymic, *nude* phenotype, opened up the possibility of genetic manipulation in the developing thymus. Re-expression of FoxN1 and the restoration of many aspects of thymic function suggested that F-box transcription factors could, as expected, direct the formation of an entire organ. The second piece of evidence implicating thymic regeneration was the observation that adult progenitor TEC (pTEC) can self-renew and give rise to mature functional TECs within the thymus and ectopically when injected into the kidney subcapsule in vivo (Rossi et al. 2006). This suggested that once a TEC progenitor cell is generated by whatever means, it could potentially give rise to all thymic components.

In seminal work, the Blackburn lab showed that FoxN1, when expressed at high levels, has the capacity to reprogram mouse embryonic fibroblasts (MEF) to a TEC-like fate (Bredenkamp et al. 2014a). After 10 days of FoxN1 expression, MEFs acquired a TEC-like morphology and began expressing genes characteristically expressed in TECs at the fetal stage. These included high levels of Dll4, SCF and CCL25 that are sufficient to support the initial stages of T cell development. As well, other hallmarks of TEC differentiation were expressed including Cytokeratin

8 and EpCAM. The expression of all of these genetic markers, as well as morphological changes were absent in control MEFs without the overexpressed FoxN1 transgene. The overexpression of FoxN1 in MEFs and their subsequent transition to TEC-like cells indicated that reprogramming to a thymic fate could occur in terminally differentiated cells. It also demonstrated that similar to Yamanaka factors, that could divert the fate of fibroblasts to become ESC-like and pluripotent (Takahashi and Yamanaka 2006), the expression of FoxN1 generated TEC-like progenitors, which were hence dubbed “induced TECs” (iTECs) (Bredenkamp et al. 2014a).

Apart from expressing key lineage-specific markers, iTECs exhibited the functional properties of *bona fide* TECs. Similar to OP9-DL cells or primary TECs that were transduced to express Dll molecules, iTECs could support T cell development even when cultured as a monolayer. Also, they responded to KGF, shown to be a TEC mitogen used after irradiation to restore thymic function (Kelly et al. 2010). Of great importance was the fact that when iTECs were co-aggregated with immature thymocytes and injected into mice in the kidney subcapsular region, iTECs were able to organize into a thymus-like structure with clearly defined medullary and cortical regions. Formation of thymus-like structures was previously possible only with ex vivo TECs derived from fetal thymus and specialized pTECs that can potentially differentiate into both cTECs and mTECs. Functionally, the graft was able to support the generation of mature SP T cells, including $CD4^+$, which OP9-DL cells fail to generate. These SP $CD8^+$ and $CD4^+$ SP cells behaved normally in their migration to the secondary lymphoid organs of spleen and lymph nodes. Altogether, these results demonstrated that expression of FoxN1 is sufficient to reprogram MEFs into TEC-like cells that are able to form functional thymic structures in vivo capable of giving rise to mature T cells.

The decline in FoxN1 expression is one of the molecular characteristics of age-related involution (Boehm and Swann 2013). Expression of FoxN1-driven genes such as Dll4 and CCL25 is also down-regulated, as is the upstream Wnt signaling pathway. There are many structural changes in the thymus including the increase in the adipose tissue and some blurring of the division of thymic medullary and cortical architecture. Attempts to reverse thymic involution include sex-steroid ablation or orchiectomy, which show some degree of success in that thymus size and thymocytes number are increased. However, these effects are transient, leaving the host with dramatic side effects after the process of atrophy resumes. In addition, the quality of the thymus did not improve in that the TECs did not display “young” TEC properties, including at the molecular level, the levels of Wnt pathway signaling molecules and FoxN1 driven genes stayed relatively the same (Griffith et al. 2012). Transgenic overexpression of FoxN1 specifically in TECs was successful in reversing thymic involution (Bredenkamp et al. 2014b). Similar to sex-steroid ablation experiments, FoxN1 overexpression in the TECs resulted in the enlargement of the thymus and an increase in T cell production. In contrast, FoxN1 overexpression led to the definitive reversing of all of the architectural and molecular hallmarks of the thymic involution back to a “young” phenotype. This included the re-establishment of cortico-medullary boundaries, the up-regulation of cell cycle molecules and an increase in expression of genes involved in T cell development (Bredenkamp et al. 2014b).

The sufficiency of FoxN1 at regenerating a complete and organized thymus and its ability to reverse thymic involution greatly advances the details of our understanding of thymus development. First, it suggests that the loss of FoxN1 expression is the key cause in thymic involution. It also indicates that involution does not result in epigenetic changes in TECs that would inhibit FoxN1 activity to reverse the tide of thymic atrophy. Because FoxN1 expression driven by a different thymic promoter did not rejuvenate the aging thymus in the same way (Zook et al. 2011), the level of FoxN1 expression likely has a very direct influence on its downstream activity. This indicated that maintenance of pTECs might be dependent on the level of FoxN1 expression. The iTEC results further support the notion that there is a common pTEC that can give rise to both cortical and medullary TECs and that its maintenance maybe key in regeneration and rejuvenation of the thymus.

In addition, there are profound clinical implications of FoxN1 in the maintenance of a healthy immune system. As discussed earlier, this approach could lead to autologous systems where matched fibroblast-derived cells could be forced to express FoxN1, rather than Dll to activate Notch (Mohtashami et al. 2013). The demonstration that FoxN1 confers the potential to form of an autologous thymus by its overexpression in fibroblasts provides an even better platform going forward. It also makes FoxN1 activation a prime endpoint for small molecule discovery for therapeutic applications in cases where the thymus has been damaged through disease or due to aging resulting in T cell depletion. The ability of a single factor to regenerate a whole organ makes it a very attractive molecule to focus on. While many but not all of the downstream targets of FoxN1 have been identified, the factors upstream of FoxN1 have not been well characterized. This additional understanding will help in strategizing for another paradigm in thymic regeneration that exploits the properties of pluripotent stem cells.

Pluripotent Stem Cells (PSC) and Thymus Regeneration

The challenge with organ regeneration is the proper integration of the graft within the body. The advantage with the thymus is the successful history of FTOC and pTEC transplant into the kidney subcapsular region. Based on their expression of EpCAM, pTECs were isolated and shown to have the capacity to differentiate into both cTEC and mTEC. This was done using $EpCAM^+$ cells injected into the kidney subcapsular region together with HSCs and fibroblasts (Rossi et al. 2006). The generation of thymus-like structures with cortical and medullary regions that could support the development of *bona fide* mature T cells confirmed the regenerative capacity of pTECs. For developmental biologists interested in regenerating a thymus, these studies identified a single cell type that could be the focus of PSC-differentiation.

The thymus can be broken down into its three components: the hematopoietic seeding cells, the epithelial TECs and the fibroblastic encasing of the thymus (Manley and Blackburn 2003). While the epithelial origin has been shown to be endodermal derived from the third pharyngeal pouch, the additional components are neural crest mesenchymal fibroblasts and the mesoderm-derived hematopoietic cells. Through a series of stepwise additions of cytokines and signaling factors, mouse ESCs or PSCs can be differentiated to form embryonic bodies followed by mesodermal or endodermal specification. A protocol based on the isolation of $EpCAM^+$ ESC-derived epithelial cells was established by culturing the embryoid bodies (Lai and Jin 2009). Upon implantation of ESC-derived pTECs under the kidney capsule, both cTECs and mTECs were formed. The anti-tumor activity of the T cells reconstituted in this model demonstrated that novel T cells were produced *de novo* and were functional and self-tolerant against both the TEC and hematopoietic components (Lai et al. 2011).

Following a similar but more regimented protocol used for mouse ESC differentiation, hESC cells were differentiated first into definitive endoderm using TGFβ and canonical WNT pathway activation (Parent et al. 2013; Sun et al. 2013). The defining endodermal transcription factor SOX17 was expressed at this stage. This was followed by addition of retinoic acid, BMP4 and a WNT inhibitor to anteriorize the developmental path towards the 3rd pharyngeal pouch, known to give rise to the thymic anlage. At this stage, the distinctive expression of HOXA3 and TBX1 was achieved. To obtain pTECs in the next stage of in vitro differentiation, FGF7, FGF8, epithelial growth factor (EGF) were added to the cultures. A strong expression of FOXN1 and PAX1 amongst other transcription factors is the most telling readout for TECs generation. However, to isolate pTECs for transplantation, the surface marker EpCAM1 was used. Once grafted, the injected pTECs continued to differentiate and began to express DLL4 and CCL25 among other factors required for T cell specification, differentiation and survival.

Transplant of human pTECs into *nude* mice also allowed the production of mouse-derived T cells. When human bone marrow derived hematopoietic progenitor cells were transplanted into these mice, they could similarly migrate to the ectopic human thymus and give rise to mature T cells. Initially, no expression of AIRE was observed in these grafts, suggesting that the T cells produced could exhibit autoimmunity (Parent et al. 2013; Sun et al. 2013). However, the graft's ability to attract hematopoietic progenitor cells was diminished after a few weeks. This suggested that an additional factor might be required to increase the longevity of ESC-derived TECs. Interestingly, when recombinant FOXN1 fused to the TAT domain of the HIV virus was added to the hESC-derived culture, it helped to sustain the pTECs (Su et al. 2015a), as TAT-fusion proteins can cross the plasma membrane barrier, increasing nuclear FOXN1 concentration and activity. FOXN1 activity appears to be key at the developmental stage when pTECs are generated. Despite only transiently adding FOXN1, the resulting TECs appear to demonstrate increased longevity when injected into mice and improved T cell production.

Protocols for obtaining all three components of the thymus (TECs, HPCs and fibroblast) from human or mouse ESCs as have been established (Kennedy et al. 2012; de Pooter et al. 2003). Human induced pluripotent stem cells (iPSCs) can be similarly differentiated to TECs and used to regenerate a thymus (Okabe et al. 2015). Although this has not yet been achieved, all three components can be potentially be derived from the same ESC or iPSC line to regenerate a whole thymus. The regeneration of a thymus using ESCs or iPSCs can provide a great advantage for genetic and immunological manipulation. Experiments harnessing this powerful tool have already begun; myelin oligodendrocyte glycoprotein overexpression in ESC-derived TECs can prevent the onset of experimental autoimmune encephalomyelitis by negatively selecting the T cells recognizing this protein (Su et al. 2015b). Other autoimmune disease phenotypes can be rescued as a consequence of the access to genetic manipulation in this system. Even more dramatic would be the transplant of histo-incompatible organs and the use of an autologous source to generate TECs to generate host and graft tolerant T cells. The main advantage of the artificial or regenerated thymus would be for it to behave like a normal thymus in reproducing a broad, yet self-tolerant T cell repertoire for transfusion/transplantation in a wide range of applications.

References

Abe N, Hozumi K, Hirano K, Yagita H, Habu S (2010) Notch ligands transduce different magnitudes of signaling critical for determination of T-cell fate. Eur J Immunol 40:2608

Anderson G, Jenkinson EJ (2008) Bringing the thymus to the bench. J Immunol 181:7435

Anderson G, Takahama Y (2012) Thymic epithelial cells: working class heroes for T cell development and repertoire selection. Trends Immunol 33:256

Anderson G, Jenkinson EJ, Moore NC, Owen JJ (1993) MHC class II-positive epithelium and mesenchyme cells are both required for T-cell development in the thymus. Nature 362:70

Armstrong F et al (2009) NOTCH is a key regulator of human T-cell acute leukemia initiating cell activity. Blood 113:1730

Artavanis-Tsakonas S, Muskavitch MA (2010) Notch: the past, the present, and the future. Curr Top Dev Biol 92:1

Awong G et al (2009) Characterization in vitro and engraftment potential in vivo of human progenitor T cells generated from hematopoietic stem cells. Blood 114:972

Awong G, Herer E, La Motte-Mohs RN, Zuniga-Pflucker JC (2011) Human CD8 T cells generated in vitro from hematopoietic stem cells are functionally mature. BMC Immunol 12:22

Awong G et al (2013) Human proT-cells generated in vitro facilitate hematopoietic stem cell-derived T-lymphopoiesis in vivo and restore thymic architecture. Blood 122:4210

Balciunaite G, Ceredig R, Fehling HJ, Zuniga-Pflucker JC, Rolink AG (2005) The role of Notch and IL-7 signaling in early thymocyte proliferation and differentiation. Eur J Immunol 35:1292

Besseyrias V et al (2007) Hierarchy of Notch-Delta interactions promoting T cell lineage commitment and maturation. J Exp Med 204:331

Blackburn CC et al (1996) The nu gene acts cell-autonomously and is required for differentiation of thymic epithelial progenitors. Proc Natl Acad Sci U S A 93:5742

Boehm T, Swann JB (2013) Thymus involution and regeneration: two sides of the same coin? Nat Rev Immunol 13:831

Bredenkamp N et al (2014a) An organized and functional thymus generated from FOXN1-reprogrammed fibroblasts. Nat Cell Biol 16:902

Bredenkamp N, Nowell CS, Blackburn CC (2014b) Regeneration of the aged thymus by a single transcription factor. Development 141:1627

Calderon L, Boehm T (2012) Synergistic, context-dependent, and hierarchical functions of epithelial components in thymic microenvironments. Cell 149:159

Coffer PJ, Burgering BM (2004) Forkhead-box transcription factors and their role in the immune system. Nat Rev Immunol 4:889

Cohen ED, Morrisey EE (2008) A house with many rooms: how the heart got its chambers with foxn4. Genes Dev 22:706

Csaszar E et al (2014) Blood stem cell fate regulation by Delta-1-mediated rewiring of IL-6 paracrine signaling. Blood 123:650

Dallas MH, Varnum-Finney B, Delaney C, Kato K, Bernstein ID (2005) Density of the Notch ligand Delta1 determines generation of B and T cell precursors from hematopoietic stem cells. J Exp Med 201:1361

de Pooter R, Zuniga-Pflucker JC (2007) T-cell potential and development in vitro: the OP9-DL1 approach. Curr Opin Immunol 19:163

de Pooter RF, Cho SK, Carlyle JR, Zuniga-Pflucker JC (2003) In vitro generation of T lymphocytes from embryonic stem cell-derived prehematopoietic progenitors. Blood 102:1649

Delaney C, Varnum-Finney B, Aoyama K, Brashem-Stein C, Bernstein ID (2005) Dose-dependent effects of the Notch ligand Delta1 on ex vivo differentiation and in vivo marrow repopulating ability of cord blood cells. Blood 106:2693

Delaney C et al (2010) Notch-mediated expansion of human cord blood progenitor cells capable of rapid myeloid reconstitution. Nat Med 16:232

Dervovic D (2011) University of Toronto

Dervovic DD, Ciofani M, Kianizad K, Zuniga-Pflucker JC (2012) Comparative and functional evaluation of in vitro generated to ex vivo CD8 T cells. J Immunol 189:3411

Dervovic DD et al (2013) Cellular and molecular requirements for the selection of in vitro-generated CD8 T cells reveal a role for Notch. J Immunol 191:1704

Dooley J, Liston A (2012) Molecular control over thymic involution: from cytokines and microRNA to aging and adipose tissue. Eur J Immunol 42:1073

Dorsch M et al (2002) Ectopic expression of Delta4 impairs hematopoietic development and leads to lymphoproliferative disease. Blood 100:2046

D'Souza B, Miyamoto A, Weinmaster G (2008) The many facets of Notch ligands. Oncogene 27:5148

Duarte A et al (2004) Dosage-sensitive requirement for mouse Dll4 in artery development. Genes Dev 18:2474

Dudakov JA et al (2012) Interleukin-22 drives endogenous thymic regeneration in mice. Science 336:91

Ebos JM et al (2009) Accelerated metastasis after short-term treatment with a potent inhibitor of tumor angiogenesis. Cancer Cell 15:232

El Kassar N et al (2004) A dose effect of IL-7 on thymocyte development. Blood 104:1419

Fares I, Rivest-Khan L, Cohen S, Sauvageau G (2015) Small molecule regulation of normal and leukemic stem cells. Curr Opin Hematol 22:309

Frank J et al (1999) Exposing the human nude phenotype. Nature 398:473

Gale NW et al (2004) Haploinsufficiency of delta-like 4 ligand results in embryonic lethality due to major defects in arterial and vascular development. Proc Natl Acad Sci U S A 101:15949

Gehre N et al (2015) A stromal cell free culture system generates mouse pro-T cells that can reconstitute T-cell compartments in vivo. Eur J Immunol 45:932

Godfrey DI, Kennedy J, Suda T, Zlotnik A (1993) A developmental pathway involving four phenotypically and functionally distinct subsets of CD3-CD4-CD8- triple-negative adult mouse thymocytes defined by CD44 and CD25 expression. J Immunol 150:4244

Goldberg GL et al (2005) Sex steroid ablation enhances lymphoid recovery following autologous hematopoietic stem cell transplantation. Transplantation 80:1604

Griffith AV, Fallahi M, Venables T, Petrie HT (2012) Persistent degenerative changes in thymic organ function revealed by an inducible model of organ regrowth. Aging Cell 11:169

Heinzel K, Benz C, Martins VC, Haidl ID, Bleul CC (2007) Bone marrow-derived hemopoietic precursors commit to the T cell lineage only after arrival in the thymic microenvironment. J Immunol 178:858

Heng TS et al (2005) Effects of castration on thymocyte development in two different models of thymic involution. J Immunol 175:2982

Holmes R, Zuniga-Pflucker JC (2009) The OP9-DL1 system: generation of T-lymphocytes from embryonic or hematopoietic stem cells in vitro. *Cold Spring Harb Protoc*, pdb prot5156

Hozumi K et al (2004) Delta-like 1 is necessary for the generation of marginal zone B cells but not T cells in vivo. Nat Immunol 5:638

Hozumi K et al (2008) Delta-like 4 is indispensable in thymic environment specific for T cell development. J Exp Med 205:2507

Hrabe de Angelis M, McIntyre J 2nd, Gossler A (1997) Maintenance of somite borders in mice requires the Delta homologue DII1. Nature 386:717

Ikawa T et al (2010) An essential developmental checkpoint for production of the T cell lineage. Science 329:93

Ito M et al (2002) NOD/SCID/gamma(c)(null) mouse: an excellent recipient mouse model for engraftment of human cells. Blood 100:3175

Jaleco AC et al (2001) Differential effects of Notch ligands Delta-1 and Jagged-1 in human lymphoid differentiation. J Exp Med 194:991

Jenkinson EJ, Owen JJ (1990) T-cell differentiation in thymus organ cultures. Semin Immunol 2:51

Jenkinson EJ, Franchi LL, Kingston R, Owen JJ (1982) Effect of deoxyguanosine on lymphopoiesis in the developing thymus rudiment in vitro: application in the production of chimeric thymus rudiments. Eur J Immunol 12:583

Jordan RK, Robinson JH, Hopkinson NA, House KC, Bentley AL (1985) Thymic epithelium and the induction of transplantation tolerance in nude mice. Nature 314:454

Katsura Y (2002) Redefinition of lymphoid progenitors. Nat Rev Immunol 2:127

Kelly RM et al (2010) Short-term inhibition of p53 combined with keratinocyte growth factor improves thymic epithelial cell recovery and enhances T-cell reconstitution after murine bone marrow transplantation. Blood 115:1088

Kennedy M et al (2012) T lymphocyte potential marks the emergence of definitive hematopoietic progenitors in human pluripotent stem cell differentiation cultures. Cell Rep 2:1722

Koch U et al (2008) Delta-like 4 is the essential, nonredundant ligand for Notch1 during thymic T cell lineage commitment. J Exp Med 205:2515

La Motte-Mohs RN, Herer E, Zuniga-Pflucker JC (2005) Induction of T-cell development from human cord blood hematopoietic stem cells by Delta-like 1 in vitro. Blood 105:1431

Lai L, Jin J (2009) Generation of thymic epithelial cell progenitors by mouse embryonic stem cells. Stem Cells 27:3012

Lai L et al (2011) Mouse embryonic stem cell-derived thymic epithelial cell progenitors enhance T-cell reconstitution after allogeneic bone marrow transplantation. Blood 118:3410

Le Borgne R, Bardin A, Schweisguth F (2005) The roles of receptor and ligand endocytosis in regulating Notch signaling. Development 132:1751

Lehar SM, Dooley J, Farr AG, Bevan MJ (2005) Notch ligands Delta 1 and Jagged1 transmit distinct signals to T-cell precursors. Blood 105:1440

Lopes N, Serge A, Ferrier P, Irla M (2015) Thymic crosstalk coordinates medulla organization and T-cell tolerance induction. Front Immunol 6:365

Love PE, Bhandoola A (2011) Signal integration and crosstalk during thymocyte migration and emigration. Nat Rev Immunol 11:469

Manley NR, Blackburn CC (2003) A developmental look at thymus organogenesis: where do the non-hematopoietic cells in the thymus come from? Curr Opin Immunol 15:225

Mohtashami M, Zuniga-Pflucker JC (2006) Three-dimensional architecture of the thymus is required to maintain delta-like expression necessary for inducing T cell development. J Immunol 176:730

Mohtashami M et al (2010) Direct comparison of Dll1- and Dll4-mediated Notch activation levels shows differential lymphomyeloid lineage commitment outcomes. J Immunol 185:867

Mohtashami M, Shah DK, Kianizad K, Awong G, Zuniga-Pflucker JC (2013) Induction of T-cell development by Delta-like 4-expressing fibroblasts. Int Immunol 25:601

Montecino-Rodriguez E, Johnson A, Dorshkind K (1996) Thymic stromal cells can support B cell differentiation from intrathymic precursors. J Immunol 156:963

Nakano T, Kodama H, Honjo T (1994) Generation of lymphohematopoietic cells from embryonic stem cells in culture. Science 265:1098

Nehls M, Pfeifer D, Schorpp M, Hedrich H, Boehm T (1994) New member of the winged-helix protein family disrupted in mouse and rat nude mutations. Nature 372:103

Nichols JT et al (2007) DSL ligand endocytosis physically dissociates Notch1 heterodimers before activating proteolysis can occur. J Cell Biol 176:445

Ohigashi I, Nitta T, Lkhagvasuren E, Yasuda H, Takahama Y (2011) Effects of RANKL on the thymic medulla. Eur J Immunol 41:1822

Okabe M, Ito S, Nishio N, Tanaka Y, Isobe K (2015) Thymic epithelial cells induced from pluripotent stem cells by a three-dimensional spheroid culture system regenerates functional T cells in nude mice. Cell Reprogram 17:368

Parent AV et al (2013) Generation of functional thymic epithelium from human embryonic stem cells that supports host T cell development. Cell Stem Cell 13:219

Preusse K et al (2015) Context-dependent functional divergence of the Notch Ligands DLL1 and DLL4 in vivo. PLoS Genet 11:e1005328

Pui JC et al (1999) Notch1 expression in early lymphopoiesis influences B versus T lineage determination. Immunity 11:299

Radtke F et al (1999) Deficient T cell fate specification in mice with an induced inactivation of Notch1. Immunity 10:547

Reimann C et al (2012) Human T-lymphoid progenitors generated in a feeder-cell-free Delta-like-4 culture system promote T-cell reconstitution in NOD/SCID/gammac(-/-) mice. Stem Cells 30:1771

Rodewald HR, Kretzschmar K, Swat W, Takeda S (1995) Intrathymically expressed c-kit ligand (stem cell factor) is a major factor driving expansion of very immature thymocytes in vivo. Immunity 3:313

Rodewald HR, Ogawa M, Haller C, Waskow C, DiSanto JP (1997) Pro-thymocyte expansion by c-kit and the common cytokine receptor gamma chain is essential for repertoire formation. Immunity 6:265

Rossi SW, Jenkinson WE, Anderson G, Jenkinson EJ (2006) Clonal analysis reveals a common progenitor for thymic cortical and medullary epithelium. Nature 441:988

Rossi SW et al (2007) RANK signals from CD4(+)3(-) inducer cells regulate development of Aire-expressing epithelial cells in the thymic medulla. J Exp Med 204:1267

Schmitt TM, Zuniga-Pflucker JC (2002) Induction of T cell development from hematopoietic progenitor cells by delta-like-1 in vitro. Immunity 17:749

Schmitt TM, Zuniga-Pflucker JC (2006) T-cell development, doing it in a dish. Immunol Rev 209:95

Schmitt TM, Ciofani M, Petrie HT, Zuniga-Pflucker JC (2004a) Maintenance of T cell specification and differentiation requires recurrent notch receptor-ligand interactions. J Exp Med 200:469

Schmitt TM et al (2004b) Induction of T cell development and establishment of T cell competence from embryonic stem cells differentiated in vitro. [see comment]. Nat Immunol 5:410

Shah DK, Zuniga-Pflucker JC (2012) Notch receptor-ligand interactions during T cell development, a ligand endocytosis-driven mechanism. Curr Top Microbiol Immunol 360:19

Shah DK, Mohtashami M, Zuniga-Pflucker JC (2012) Role of recycling, mindbomb1 association, and exclusion from lipid rafts of delta-like 4 for effective notch signaling to drive T cell development. J Immunol
Smeets MF et al (2013) Removal of myeloid cytokines from the cellular environment enhances T-cell development in vitro. Int Immunol 25:589
Su M et al (2015a) Efficient in vitro generation of functional thymic epithelial progenitors from human embryonic stem cells. Sci Rep 5:9882
Su M et al (2015b) Administration of embryonic stem cell-derived thymic epithelial progenitors expressing MOG induces antigen-specific tolerance and ameliorates experimental autoimmune encephalomyelitis. J Autoimmun 58:36
Sun X et al (2013) Directed differentiation of human embryonic stem cells into thymic epithelial progenitor-like cells reconstitutes the thymic microenvironment in vivo. Cell Stem Cell 13:230
Swann JB et al (2014) Conversion of the thymus into a bipotent lymphoid organ by replacement of FOXN1 with its paralog, FOXN4. Cell Rep 8:1184
Swiatek PJ, Lindsell CE, del Amo FF, Weinmaster G, Gridley T (1994) Notch1 is essential for postimplantation development in mice. Genes Dev 8:707
Takahama Y (2006) Journey through the thymus: stromal guides for T-cell development and selection. Nat Rev Immunol 6:127
Takahashi K, Yamanaka S (2006) Induction of pluripotent stem cells from mouse embryonic and adult fibroblast cultures by defined factors. Cell 126:663
Tanigaki K, Honjo T (2007) Regulation of lymphocyte development by Notch signaling. Nat Immunol 8:451
Timmermans F et al (2009) Generation of T cells from human embryonic stem cell-derived hematopoietic zones. J Immunol 182:6879
Tsukamoto N, Itoi M, Nishikawa M, Amagai T (2005) Lack of Delta like 1 and 4 expressions in nude thymus anlages. Cell Immunol 234:77
Van Coppernolle S et al (2009) Functionally mature CD4 and CD8 TCRalphabeta cells are generated in OP9-DL1 cultures from human CD34+ hematopoietic cells. J Immunol 183:4859
Varnum-Finney B, et al. (2000) Immobilization of Notch ligand, Delta-1, is required for induction of notch signaling. *J Cell Sci* 113, Pt 23, 4313
Vodyanik MA, Thomson JA, Slukvin II (2006) Leukosialin (CD43) defines hematopoietic progenitors in human embryonic stem cell differentiation cultures. Blood 108:2095
von Freeden-Jeffry U et al (1995) Lymphopenia in interleukin (IL)-7 gene-deleted mice identifies IL-7 as a nonredundant cytokine. J Exp Med 181:1519
Wada H et al (2008) Adult T-cell progenitors retain myeloid potential. Nature 452:768
Wang H, Pierce LJ, Spangrude GJ (2006) Distinct roles of IL-7 and stem cell factor in the OP9-DL1 T-cell differentiation culture system. Exp Hematol 34:1730
Watanabe Y, Gyotoku J, Katsura Y (1989) Analysis of the development of T cells by transferring precursors into cultured fetal thymus with a microinjector. Thymus 13:57
Zakrzewski JL et al (2006) Adoptive transfer of T-cell precursors enhances T-cell reconstitution after allogeneic hematopoietic stem cell transplantation. Nat Med 12:1039
Zakrzewski JL et al (2008) Tumor immunotherapy across MHC barriers using allogeneic T-cell precursors. Nat Biotechnol 26:453
Zediak VP, Bhandoola A (2005) Aging and T cell development: interplay between progenitors and their environment. Semin Immunol 17:337
Zook EC et al (2011) Overexpression of Foxn1 attenuates age-associated thymic involution and prevents the expansion of peripheral CD4 memory T cells. Blood 118:5723
Zuniga-Pflucker JC (2004) T-cell development made simple. Nat Rev Immunol 4:67

Chapter 5
Spleen Development and the Origin of Stromal Diversity

Elisa Lenti and Andrea Brendolan

Abstract The mammalian spleen is a secondary lymphoid organ that plays an important role in hematopoiesis and host defense. Consequently, asplenic or post-splenectomised patients have often increased risk of overwhelming infections as compared to normal population. Although much remains to be elucidated to fully understand the mechanisms governing spleen development, work over the past decade has provided new insights into the cellular and molecular mechanisms involved. It is now clear that spleen development requires coordination of cell fate specification, migration and differentiation, and failure to properly accomplish these stepwise processes results in spleen growth defects. Recent work has also provided new insights on the origin of mature spleen stromal cells by demonstrating the developmental relationship between the embryonic splenic mesenchyme and adult stromal microenvironment. These findings established that, with the exception of endothelial cells, all mature spleen stromal cells originate from embryonic multipotent mesenchymal precursors. Here we summarize the current knowledge on the key transcription factors involved in patterning and morphogenesis of the mouse spleen, and describe the developmental origin of spleen stromal cells.

List of Abbreviations

SMP	Splanchnic mesodermal plate
SPM	Spleno-pancreatic mesenchyme
DP	Dorsal pancreas
PBX1	Pre-B-Cell Leukemia Homeobox 1
NKX2-5	NK2 Homeobox 5
BAPX1	Bagpipe Homeobox Protein Homolog 1
SOX9	Sry-related hmg-box 9
TLX1	T-Cell Leukemia Homeobox 1
WT1	Wilms Tumor 1

E. Lenti • A. Brendolan (✉)
Division of Experimental Oncology, IRCCS San Raffaele Scientific Institute, Milan, Italy
e-mail: brendolan.andrea@hsr.it

T. Watanabe, Y. Takahama (eds.), *Synthetic Immunology*,
DOI 10.1007/978-4-431-56027-2_5

SF1	Steoidogenic
NKX2-3	NK2 Homeobox 3
POD1	Podocyte-Expressed 1
RPSA	Ribosomal Protein SA
CDKN2B	Cyclin-dependent kinase 4 inhibitor B
VE-cadherin	Vascular endothelial cadherin
MAdCAM1	Mucosal vascular addressin cell adhesion molecule 1
PDGFRβ	Platelet-Derived Growth Factor Receptor, Beta
ICA	Isolated congenital asplenia
CXCL12	C-X-C motif chemokine 12, also known as stromal cell-derived factor 1 (SDF-1)
CXCL13	C-X-C motif chemokine 13, also known as B lymphocyte chemoattractant (BLC)
CCL19	C-C motif chemokine 19
CCL21	C-C motif chemokine 21
RORγt	Retinoic acid-related orphan receptor-γ t
LTβR	Lymphotoxin beta receptor
LTα1β2	Lymphotoxin-α1β2
LTi	Lymphoid tissue inducer cells
LTo	Lymphoid tissue organizer cells
LCMV	Lymphocytic choriomeningiatis virus
FRC	Fibroblastic reticular cell
FDC	Follicular dendritic cell
MRC	Marginal reticular cell
ECM	Extracellular matrix
PALS	Periasteriolar lymphoid sheet
BCR	B-cell receptor
ALO	Artificial lymphoid organs

Spleen Morphogenesis

Spleen organogenesis forms as a result of a complex interplay of molecular and cellular networks (Brendolan et al. 2007; Burn et al. 2008). Recent work has provided the first comprehensive analysis of the morphogenetic events occurring during early spleen formation (Burn et al. 2008). These findings showed that at E9.5 the splanchnic mesodermal plate (SMP), a thick epithelial-like plate of cells present on each side of the embryonic gut surrounding the endoderm expresses the spleen-specific genes *Bapx1* and *Nkx2-5*, bilaterally (Hecksher-Sorensen et al. 2004). However, by E10.5 the SMP is present only on the left-hand side accompanying the leftward growth in the dorsal pancreatic mesenchyme (Fig. 5.1). At this stage, the expression of *Bapx1* and *Nkx2-5* appears in different domains of the dorsal pancreatic mesenchyme, though in situ hybridization analyses revealed that newly specified spleen mesenchymal cells expressing a defined set of

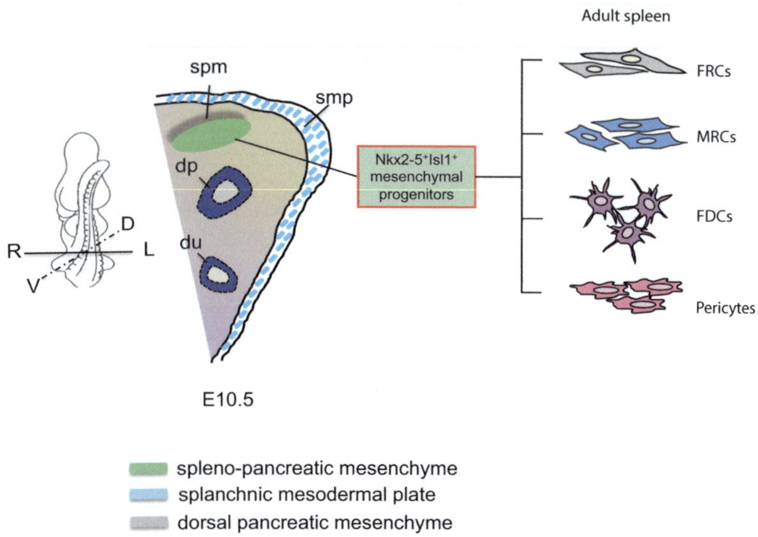

Fig. 5.1 Developmental relationship between embryonic spleen mesenchymal cells and adult stromal cell subsets. At E10.5, the SMP is present only on the left-hand side accompanying the leftward growth in the dorsal pancreas. Spleen mesenchymal cells are specified at this stage and localize on the dorsal domain of the pancreatic mesenchyme (green area). Newly specified mesenchymal progenitors are characterized by the expression of *Nkx2-5* and *Isl1* and are responsible to generate all mature spleen stromal cell subsets. smp, splanchnic mesodermal plate; spm, spleno-pancreatic mesenchyme; dp, dorsal pancreas; du, duodenum; FDCs, follicular dendritic cells; FRCs, fibroblastic reticular cells; MRCs, marginal reticular cells

transcription factors localize specifically in the dorsal domain (Hecksher-Sorensen et al. 2004). Supporting a role for the dorsal pancreas, as the source of spleen mesenchymal progenitors is the recent demonstration that ablation of the dorsal pancreatic mesenchyme results in agenesis of both pancreas and spleen (Landsman et al. 2011). At E10.5, the spleno-pancreatic mesenchyme is still surrounded by SMP whose role in spleen formation remains unknown (Fig. 5.1). Interestingly, the finding that embryos heterozygous for the dominant hemimelia (Dh) mutation do not have the SMP (Green 1967) and are born asplenic, while their pancreas appears normal, suggests that the SMP may act as a signaling center for the underlying splenic mesenchyme (Hecksher-Sorensen et al. 2004). At present, the signals emanating from the SMP that are required for spleen development remains unknown, as it is the Dh mutation.

Once specified at E10.5, spleen mesenchymal cells expand along the posterior mesenchyme that overlies the left hand side of the stomach towards the anterior portion, and at E12 mesenchymal cells condense within the dorsal mesogastrium adjacent to the stomach and dorsal pancreas to form a discrete splenic primordium (Burn et al. 2008; Patterson et al. 2000). At this stage, the developing spleen appears as a distinct tissue separated from the neighbouring dorsal pancreas and stomach. It has been demonstrated that the homeodomain transcription factor BAPX1 plays a critical role in controlling the position and separation of the spleen from the adjacent dorsal pancreas (Hecksher-Sorensen et al. 2004; Tribioli and Lufkin 1999), and that in the absence of *Bapx1* the putative splenic mesenchyme remains attached to the dorsal pancreas and spleen fails to form as early as E11 (Lettice et al. 1999). By E13 onward, the active proliferation of mesenchymal and endothelial cells together with the colonization of hematopoietic precursors support the second phase of spleen development, which is characterized by organ expansion and patterning of lymphoid compartments (Godin et al. 1999).

Genetic and Transcriptional Control of Spleen Mesenchymal Cell Specification

Initiation of spleen development consists in the specification of mesenchymal progenitors at E10.5 within the dorsal pancreatic mesenchyme (Brendolan et al. 2005; Hecksher-Sorensen et al. 2004). Analysis of knockout mouse models revealed the critical role played by a set of genes, mainly encoding transcription factors, in activating spleen-specific genetic programs involved in cell-fate specification, cell proliferation, and differentiation (Table 5.1) (Biben et al. 2000; Dear et al. 1995; Harvey 2002; Herzer et al. 1999; Kanzler and Dear 2001; Lettice

Table 5.1 Mouse mutants displaying defects of spleen development

Gene Name	Spleen Phenotype	Other abnormalities	Splenic defect
Tlx1 (Hox11)	Agenesis	No other abnormalities	Impaired cell-fate specification, reduced proliferation and organ expansion.
Pbx1	Agenesis	Abnormal development or agenesis of most organ systems, lethal (E15.5)	Impaired cell-fate specification, reduced proliferation and organ expansion.
Wt1	Hypoplasia	Failure of kidney, heart, diaphragm, gonad formation, lethal (E15)	Increased apoptosis and organ expansion
Nkx2-5	Hypoplasia	Failure of heart formation, lethal (E9.5-10)	Unknown
Nkx2-3	Agenesis or hypoplasia	Small intestine defects	Impaired architecture and lymphoid composition
Bapx1 (Nkx3-2)	Agenesis	Abnormal development of small intestine, skeletal defects	Impaired cell-fate specification and morphogenesis
Pod1/ Tcf21 (Capsulin)	Agenesis	Lung and kidney defects	Increased apoptosis and organ expansion
Sox11	Agenesis	Heart, lungs, stomach, pancreas, skeleton defects, perinatal lethal	Impaired organ expansion
Sf1 (NR5a1)	Hypoplasia	Adrenal glands, gonads, sex determination defects	Impaired architecture and organ expansion

et al. 1999; Lu et al. 2000; Lyons et al. 1995; Prall et al. 2007; Quaggin et al. 1999; Roberts et al. 1995; Selleri et al. 2001). At E10.5 fully specified spleen mesenchymal progenitors are characterized by the expression of *Bapx1*, *Pbx1*, *Tlx1*, *Wt1*, *Nkx2-5*, *Sox11* and *Pod1*. At present, it remains unknown whether *Nkx2-3* and *Sf1* are expressed at this early stage. However, the expression of *Bapx1, Pbx1 and Nkx2-5* occurs slightly before *Tlx1* (Brendolan, unpublished observation). At E12.5, assessment of gene expression in different mutant embryos by in situ hybridization analysis revealed a hierarchical relationship whereby *Bapx1, Pbx1* and *Pod1* act upstream of *Tlx1*, *Wt1* and *Nkx2-5* (Brendolan et al. 2005). At the transcriptional level, it was shown that PBX1 directly binds to *Tlx1* and *Nkx2-5* promoters (Koss et al. 2012). Importantly, while PBX1 is required to initiate *Tlx1* and *Nkx2-5* expression in spleen mesenchymal cells, ablation of *Pbx1* after specification of progenitor cells revealed a role of PBX1 in maintaining *Nkx2-5* but not *Tlx1* expression (Koss et al. 2012). In addition, it was shown that TLX1 directly controls *Wt1* transcription, as *Wt1* mRNA expression is lost in the splenic anlage of *Tlx1* mutant embryos (Koehler et al. 2000). Furthermore, previous work showed that *Nkx2-5* expression is maintained in *Tlx1* mutants from a CD1 genetic background. However, recent findings from our laboratory revealed that expression of *Nkx2-5* is severely reduced in *Tlx1* mutant embryonic spleens of mice carrying a C57BL6 genetic background (Lenti, unpublished data). These findings indicate that *Tlx1* may act upstream of *Nkx2-5* depending on the genetic background. Whether BAPX1 and POD1 directly control the expression of *Tlx1* and *Nkx2-5* remain to be determined. Although the mechanism responsible for the spleen defect of *Sox11* mutants has not been elucidated, it was shown that initiation of spleen development in the absence of *Sox11* commences normally as E12.5 mutants have a normal splenic anlage (Hornblad et al. 2011). *Sf1* mutants have also severe spleen hypoplasia; though the role of SF1 during initiation of spleen development remains unclear.

Analysis of knockout mice for the aforementioned transcription factors revealed that only *Tlx1* mutants show asplenia as the sole organ abnormality (Dear et al. 1995; Roberts et al. 1995). Conversely, knockout for all the other transcription factor-encoding genes have multiple abnormalities in addition to asplenia or spleen hypoplasia (Brendolan et al. 2007). Thus, the phenotype of *Tlx1* mutants is the only that resembles human isolated congenital asplenia (ICA), a condition in which the absence of the spleen exists as the only abnormality (Koss et al. 2012; Mahlaoui et al. 2011). At present, only heterozygous mutations of the Ribosomal Protein SA (*RPSA*) gene were found associated to ICA in humans (Bolze et al. 2013). However, given that *Rpsa* homozygous mutants are embryonic lethal, whereas heterozygous mice have a normal spleen, the precise role of *Rpsa* during mouse spleen development remains unclear (Bolze et al. 2013).

Absence or impaired spleen development has been associated with distinct cellular abnormalities. Indeed, while increased apoptosis was found in the splenic mesenchyme of *Pod1* or *Wt1* mutants, reduced mesenchymal proliferation was detected in the absence of *Pbx1* or *Tlx1* (Brendolan et al. 2005; Herzer et al. 1999; Quaggin et al. 1999). Interestingly, *Pbx1* and *Tlx1* genetically interact as double compound mice display smaller spleen as compared to single heterozygous controls (Brendolan et al. 2005). Similarly, *Tlx1* and *Nkx2-5* compound mice display also smaller spleens as compared to controls (Lenti, unpublished data). During spleen development, PBX1 controls cellular proliferation by repressing *Cdkn2b*, a CDK inhibitor overexpressed in the absence of *Pbx1* (Koss et al. 2012; Miller et al. 2007). *Nkx2-5* heterozygous embryos have also spleen hypoplasia, though it is unclear whether this phenotype results from reduced proliferation of spleen mesenchymal progenitors. Nevertheless, the finding that NKX2-5 also binds the *Cdkn2b* promoter suggests that this transcription factor may regulate cellular proliferation during spleen organogenesis (Koss et al. 2012). Overall, these findings indicate a crucial role for PBX1, TLX1 and NKX2-5 in regulating proliferation of spleen mesenchymal progenitors, though other pathways likely contribute to the spleen defects observed in these mutants. The identification of downstream target genes controlled by the aforementioned transcription factors will contribute to better elucidate the mechanism by which they promote spleen development.

Hematopoietic-Cell Colonization and the Establishment of Lymphoid Compartments

At E13, the spleen is colonized by different hematopoietic progenitors emigrating from the foetal liver. CD45+ cells expand rapidly and by E15.5 they represent more than 60 % of the total cells (Vondenhoff et al. 2009). At this stage, the embryonic spleen is enriched for a $CD4^{int}cKit^{lo}Lin^{-}$ population mostly committed to a lymphoid fate and capable to differentiate into B, T, NK and, to a lesser extent, myeloid cells (Desanti et al. 2008; Vondenhoff et al. 2009). Among those hematopoietic progenitors, lymphoid tissue inducer (LTi) cells are also present and distributed in the proximity of blood vessels, although some LTi cells are also scattered throughout the embryonic spleen. From E16.5 onward, LTi cells organize in a ring-like pattern around $VE\text{-}Cadherin^{+}MAdCAM\text{-}1^{+}$ endothelial cells representing near 20 % of the total CD45- stromal population. Analysis of distribution patter of lymphoid chemokines revealed that CXCL12 and CXCL13 were similarly expressed in the endothelial and stromal cells. On the contrary, the expression of CCL19 and CCL21 was hundred to thousand folds higher in the endothelial as compared to the stromal fraction. The peculiar distribution of LTi cells surrounding VE-Cadherin + endothelial cells and stromal progenitors coincides with the formation of the white pulp compartment (Desanti et al. 2008; Withers et al. 2007). Indeed, at E15.5 LTi and B cells segregate in regions different from those containing Ter119+ erythrocytes normally found in the red pulp, indicating that patterning of the white and red pulp areas already occurs during embryonic development.

Whereas LN LTi cells express high levels of lymphotoxin (LT) $\alpha 1\beta 2$, in the spleen only a small percentage of LTi cells express $LT\alpha 1\beta 2$, indicating that the establishment of the white pulp areas during embryonic development does not depend on this signaling pathway (Desanti et al. 2008; Withers et al. 2007). This finding is also in line with the fact that, contrary to LN formation that requires interactions between LTi and lymphoid tissue organizer (LTo) cells of mesenchymal origin, spleen organogenesis and segregation of T and B cell zones occur normally in mice deficient for RORγt or Id2 and lacking LTi cells. These findings indicate that the interactions between LTi cells and stromal progenitors through LT signaling do not appear to play a critical role for spleen development (Eberl et al. 2004; Yokota et al. 1999). In agreement with this, ablation of LT signalling in newborn mice, but not at postnatal day 7, causes white pulp formation defects. A requirement for early LT signaling has been also demonstrated in a recent mouse model of spleen neogenesis. It was shown that early postnatal priming of mesenchymal stromal cells by LT signaling was sufficient to promote spleen formation (Tan and Watanabe 2014). Overall, these findings indicate a different requirement for LT signaling during embryonic and postnatal spleen development. Nevertheless, the postnatal organization of the white pulp requires $LT\alpha1\beta2$, and LTi cells may contribute to this process (Ngo et al. 2001; Withers et al. 2007).

Spleen Stromal Cells and the Origin of Stromal Diversity

The composition of the stromal microenvironment of the splenic white pulp resembles that of other secondary lymphoid tissues such as LNs. During postnatal life, spleen architecture is orchestrated and maintained by the activity of distinct stromal cell subsets namely fibroblastic reticular cells (FRCs) of the T-zone, follicular dendritic cells (FDCs) of the B-cell follicle and marginal reticular cells (MRCs) localized beneath the marginal sinus in the outermost part of the B-cell zone (den Haan et al. 2012; Mueller and Germain 2009). These cells express receptors of LT system including LTβR, that upon engagement with their ligands (e.g. $LT\alpha1\beta2$) expressed on lymphoid cells trigger the secretion of CCL19/CCL21 and CXCL13 lymphoid chemokines. These molecules are required for attracting T- and B- cells and organizing the lymphoid compartments (Lu and Browning 2014; Mueller and Germain 2009). Indeed mice deficient for LTβR for genes encoding lymphoid chemokines secreted by stromal cells have profound disorganization of the white pulp area and defective immune functions (Mackay et al. 1997), demonstrating the critical role of stromal cells as "organizers" of the lymphoid compartments. Stromal cells also produce the extracellular matrix (ECM), a structural scaffold of reticular fibers composed of basement membrane and interstitial matrix components (Lokmic et al. 2008). In the T-cell zone FRCs form the so-called conduit system, a reticular collagenous network that allows the transport and distribution of small molecules or particles of approximately 70 KDa from the blood to the white pulp (Mebius and Kraal 2005). The conduit also acts as path guiding T cell entry from the marginal zone bridging channels into the periarteriolar lymphoid sheath (PALS). Attraction of T cells into the PALS is promoted by CCL19 and CCL21 expressed by FRC, and CCL19 is also important to maintain T-cell homeostasis (Mueller and Germain 2009). In the B-cell follicle, FDCs play a crucial role in promoting B-cell immunity (Allen and Cyster 2008). FDCs promote recruitment of B-lymphocytes into the follicles through secretion of CXCL13 that binds CXCR5 expressed on B cells. This stromal cell type traps and presents antigens in the form of immune complexes (IC) via Fc and complement receptors, thus stimulating B cells through the B-cell receptor (BCR) and promoting germinal center formation. Generation of FDC networks relies on TNF and LT signalings; however, only LT signaling was shown to be required for FDC maintenance (Aguzzi et al. 2014). MRCs are stromal cells that localize in the outer region of the B-cell follicle and express CXCL13 and MAdCAM-1 (Katakai 2012). Although the function of MRCs remains elusive, the expression of B-cell chemokines and the close association of this cell type with CD169+ marginal metallophilic macrophages suggest a possible involvement of MRCs in supporting local niches in the outer follicular regions and marginal zone. In addition to the white pulp, the spleen contains the red pulp, a structure that plays an important role in filtering the blood. The red pulp comprises of a dense reticular network of fibroblasts and collagenous fibers intermingled with F4/80+ macrophages that contribute to support some of the functions exerted by the red pulp, including iron recycling and immune defense (Mebius and Kraal 2005; Mueller and Germain 2009). Contrary to white pulp stromal cells, the origin and function of stromal cells within the red pulp remain mostly elusive.

The origin of stromal cells in peripheral lymphoid tissues has been questioned for long time and, only recently, lineage-tracing analysis has shed light on their ontogeny. By genetic labeling of spleen mesenchymal cells within the nascent splenic anlagen, it was demonstrated the developmental relationship between embryonic mesenchymal and mature stromal cells (Boehm 2013; Castagnaro et al. 2013). These findings established that, with the exception of endothelial cells, all mature spleen stromal cells originate from embryonic multipotent mesenchymal precursors of the Nkx2-5^{+}Isl1^{+} lineage (Castagnaro et al. 2013) (Fig. 5.1). This lineage also includes lymphoid tissue organizer cells as demonstrated by the capacity of embryonic Nkx2-5^{+}Isl1^{+} mesenchymal cells transplanted into collagenous scaffolds to promote the formation of artificial lymphoid-like structures. During the course of lymphocytic choriomeningitis virus (LCMV) infection, the splenic stromal network is partially destroyed due to the cytotoxic activity of anti LCMV-specific CD8 T cells against infected stromal cells. Still, when the infection is cleared, restoration of stromal architecture occurs rapidly and within four weeks post infection (Scandella et al. 2008). It was shown that restoration of stromal integrity in LCMV-infected mice is delayed in the absence of LTi cells and LTβR engagement (Scandella et al. 2008). Furthermore, using LCMV infection as a model to study spleen regeneration, it was shown that resident stromal cells of the Nkx2-5^{+}Isl1^{+} lineage contribute to restore the stromal network integrity (Castagnaro et al. 2013). Thus, the formation and regeneration of the spleen stromal network depend exclusively on the activity of embryonic mesenchymal descendants of the Nkx2-5^{+}Isl1^{+} lineage. Importantly, regeneration of the stromal network occurs through the expansion of resident stromal cells, indicating that the spleen harbors a population of cells with regenerative capacity and possible stem cell activity (Castagnaro et al. 2013). Interestingly, this lineage does not generate the stromal compartment of other peripheral lymphoid tissues such as LNs, suggesting that tissue specific lineages contribute to generate the stromal diversity of peripheral lymphoid tissues. Additional genetic labeling experiments also demonstrated that spleen FDCs arise from perivascular platelet-derived growth factor receptor beta (PDGFRβ) positive mesenchymal precursors (Krautler et al. 2012). Thus, it appears that Nkx2-5^{+}Isl1^{+} descendants located perivascularly and expressing PDGFRβ generate FDCs of the spleen.

From Spleen Development to the Generation of Artificial Lymphoid Tissues

Work over the last decade demonstrated the critical role of embryonic spleen mesenchymal cells and mature stromal elements in regulating spleen organogenesis, immune functions and tissue regeneration. As described, the formation of distinct splenic compartments and the acquisition of immune functions require sequential steps. Indeed, (i) specification of early mesenchymal progenitors is the first critical step leading to the formation of the splenic anlage; (ii) colonization of the splenic primordium by hematopoietic progenitors and the establishment of lympho-mesenchymal interactions is the second important event underlying patterning of the splenic white pulp; concomitantly, (iii) the postnatal differentiation of embryonic mesenchymal cells into distinct stromal cell subsets by the LT signaling pathway is the last crucial event leading to the acquisition of mature stromal cells and functional lymphoid compartments. A complete understanding of how these sequential events are executed will provide important insights for better elucidating the mechanisms underlying asplenia. Furthermore, uncovering how these processes are orchestrated will provide the basic knowledge for developing strategies aiming to generate artificial lymphoid organs (aLOs) for clinical applications. Toward this end, work over the past decade has demonstrated the possibility to develop functional aLOs using basic components of lymphoid tissues such as stromal cells and lymphoid chemokines (Kobayashi and Watanabe 2010; Suematsu and Watanabe 2004). Indeed, the recapitulation of organogenetic processes by the use of defined cell types and factors will be important to develop next-generation aLOs. Spleen-specific transcription factors could be used to specify uncommitted mesodermal progenitors – derived from pluripotent stem cells – into stromal precursors. Similarly, signals that induce the formation of lymphoid compartments and maturation of stromal cells, including LT could be used to promote the differentiation of newly specified stromal progenitors isolated from pluripotent stem cells. Additionally, aLOs could be used to manipulate the immune system or restore immune functions in different context including cancer, autoimmunity and infections (Cupedo et al. 2012; Kobayashi et al. 2011). Finally, they may serve as devices to replace lymphoid tissues damaged by chemo or radiation therapy or resected due to metastatic cancer in order to avoid the development of secondary lymphedema (Tammela et al. 2007).

Concluding Remarks

The mammalian spleen plays a critical role in hematopoiesis and host defense. Although much has been done to elucidate the mechanisms underlying spleen development, there is still much to do in order to fully understand the mechanisms underlying development and organization of lymphoid compartments, and the acquisition of stromal identity. What are the signaling pathways involved in patterning the white and red pulp during embryonic development? Which are the transcription factors and developmental pathways involved in the differentiation of stromal progenitors and acquisition of stromal identify. These are only some of the questions that still wait to be answered.

Acknowledgements This was supported by funding from the Italian Association for Cancer Research (Grant IG # 14511) and the Italian Ministry of Health (RF-2011-02347691) to A.B.

References

Aguzzi A, Kranich J, Krautler NJ (2014) Follicular dendritic cells: origin, phenotype, and function in health and disease. Trends Immunol 35(3):105–113. doi:10.1016/j.it.2013.11.001

Allen CD, Cyster JG (2008) Follicular dendritic cell networks of primary follicles and germinal centers: phenotype and function. Semin Immunol 20(1):14–25. doi:10.1016/j.smim.2007.12.001

Biben C, Weber R, Kesteven S, Stanley E, McDonald L, Elliott DA, Barnett L, Koentgen F, Robb L, Feneley M, Harvey RP (2000) Cardiac septal and valvular dysmorphogenesis in mice heterozygous for mutations in the homeobox gene Nkx2-5. Circ Res 87(10):888–895

Boehm T (2013) Same function, different origins: multipotent stromal precursors in lymphoid tissues. Cell Stem Cell 12(5):501–503. doi:10.1016/j.stem.2013.04.011

Bolze A, Mahlaoui N, Byun M, Turner B, Trede N, Ellis SR, Abhyankar A, Itan Y, Patin E, Brebner S, Sackstein P, Puel A, Picard C, Abel L, Quintana-Murci L, Faust SN, Williams AP, Baretto R, Duddridge M, Kini U, Pollard AJ, Gaud C, Frange P, Orbach D, Emile JF, Stephan JL, Sorensen R, Plebani A, Hammarstrom L, Conley ME, Selleri L, Casanova JL (2013) Ribosomal protein SA haploinsufficiency in humans with isolated congenital asplenia. Science 340(6135):976–978. doi:10.1126/science.1234864

Brendolan A, Ferretti E, Salsi V, Moses K, Quaggin S, Blasi F, Cleary ML, Selleri L (2005) A Pbx1-dependent genetic and transcriptional network regulates spleen ontogeny. Development 132(13):3113–3126. doi:10.1242/dev.01884

Brendolan A, Rosado MM, Carsetti R, Selleri L, Dear TN (2007) Development and function of the mammalian spleen. BioEssays 29(2):166–177. doi:10.1002/bies.20528

Burn SF, Boot MJ, de Angelis C, Doohan R, Arques CG, Torres M, Hill RE (2008) The dynamics of spleen morphogenesis. Dev Biol 318(2):303–311. doi:10.1016/j.ydbio.2008.03.031

Castagnaro L, Lenti E, Maruzzelli S, Spinardi L, Migliori E, Farinello D, Sitia G, Harrelson Z, Evans SM, Guidotti LG, Harvey RP, Brendolan A (2013) Nkx2-5(+)islet1(+) mesenchymal

precursors generate distinct spleen stromal cell subsets and participate in restoring stromal network integrity. Immunity 38(4):782–791. doi:10.1016/j.immuni.2012.12.005
Cupedo T, Stroock A, Coles M (2012) Application of tissue engineering to the immune system: development of artificial lymph nodes. Front Immunol 3:343. doi:10.3389/fimmu.2012.00343
Dear TN, Colledge WH, Carlton MB, Lavenir I, Larson T, Smith AJ, Warren AJ, Evans MJ, Sofroniew MV, Rabbitts TH (1995) The Hox11 gene is essential for cell survival during spleen development. Development 121(9):2909–2915
den Haan JM, Mebius RE, Kraal G (2012) Stromal cells of the mouse spleen. Front Immunol 3:201. doi:10.3389/fimmu.2012.00201
Desanti GE, Cumano A, Golub R (2008) Identification of CD4int progenitors in mouse fetal spleen, a source of resident lymphoid cells. J Leukoc Biol 83(5):1145–1154. doi:10.1189/jlb.1107755
Eberl G, Marmon S, Sunshine MJ, Rennert PD, Choi Y, Littman DR (2004) An essential function for the nuclear receptor RORgamma(t) in the generation of fetal lymphoid tissue inducer cells. Nat Immunol 5(1):64–73. doi:10.1038/ni1022
Godin I, Garcia-Porrero JA, Dieterlen-Lievre F, Cumano A (1999) Stem cell emergence and hemopoietic activity are incompatible in mouse intraembryonic sites. J Exp Med 190(1):43–52
Green MC (1967) A defect of the splanchnic mesoderm caused by the mutant gene dominant hemimelia in the mouse. Dev Biol 15(1):62–89
Harvey RP (2002) Patterning the vertebrate heart. Nat Rev Genet 3(7):544–556. doi:10.1038/nrg843
Hecksher-Sorensen J, Watson RP, Lettice LA, Serup P, Eley L, De Angelis C, Ahlgren U, Hill RE (2004) The splanchnic mesodermal plate directs spleen and pancreatic laterality, and is regulated by Bapx1/Nkx3.2. Development 131(19):4665–4675. doi:10.1242/dev.01364
Herzer U, Crocoll A, Barton D, Howells N, Englert C (1999) The Wilms tumor suppressor gene wt1 is required for development of the spleen. Curr Biol 9(15):837–840
Hornblad A, Eriksson AU, Sock E, Hill RE, Ahlgren U (2011) Impaired spleen formation perturbs morphogenesis of the gastric lobe of the pancreas. PLoS ONE 6(6), e21753. doi:10.1371/journal.pone.0021753
Kanzler B, Dear TN (2001) Hox11 acts cell autonomously in spleen development and its absence results in altered cell fate of mesenchymal spleen precursors. Dev Biol 234(1):231–243. doi:10.1006/dbio.2001.0239
Katakai T (2012) Marginal reticular cells: a stromal subset directly descended from the lymphoid tissue organizer. Front Immunol 3:200. doi:10.3389/fimmu.2012.00200
Kobayashi Y, Watanabe T (2010) Synthesis of artificial lymphoid tissue with immunological function. Trends Immunol 31(11):422–428. doi:10.1016/j.it.2010.09.002
Kobayashi Y, Kato K, Watanabe T (2011) Synthesis of functional artificial lymphoid tissues. Discov Med 12(65):351–362
Koehler K, Franz T, Dear TN (2000) Hox11 is required to maintain normal Wt1 mRNA levels in the developing spleen. Dev Dyn 218(1):201–206. doi:10.1002/(SICI)1097-0177(200005)218:1<201::AID-DVDY18>3.0.CO;2-R
Koss M, Bolze A, Brendolan A, Saggese M, Capellini TD, Bojilova E, Boisson B, Prall OW, Elliott DA, Solloway M, Lenti E, Hidaka C, Chang CP, Mahlaoui N, Harvey RP, Casanova JL, Selleri L (2012) Congenital asplenia in mice and humans with mutations in a Pbx/Nkx2-5/p15 module. Dev Cell 22(5):913–926. doi:10.1016/j.devcel.2012.02.009
Krautler NJ, Kana V, Kranich J, Tian Y, Perera D, Lemm D, Schwarz P, Armulik A, Browning JL, Tallquist M, Buch T, Oliveira-Martins JB, Zhu C, Hermann M, Wagner U, Brink R, Heikenwalder M, Aguzzi A (2012) Follicular dendritic cells emerge from ubiquitous perivascular precursors. Cell 150(1):194–206. doi:10.1016/j.cell.2012.05.032
Landsman L, Nijagal A, Whitchurch TJ, Vanderlaan RL, Zimmer WE, Mackenzie TC, Hebrok M (2011) Pancreatic mesenchyme regulates epithelial organogenesis throughout development. PLoS Biol 9(9):e1001143. doi:10.1371/journal.pbio.1001143

Lettice LA, Purdie LA, Carlson GJ, Kilanowski F, Dorin J, Hill RE (1999) The mouse bagpipe gene controls development of axial skeleton, skull, and spleen. Proc Natl Acad Sci U S A 96 (17):9695–9700

Lokmic Z, Lammermann T, Sixt M, Cardell S, Hallmann R, Sorokin L (2008) The extracellular matrix of the spleen as a potential organizer of immune cell compartments. Semin Immunol 20 (1):4–13. doi:10.1016/j.smim.2007.12.009

Lu TT, Browning JL (2014) Role of the Lymphotoxin/LIGHT System in the Development and Maintenance of Reticular Networks and Vasculature in Lymphoid Tissues. Front Immunol 5:47. doi:10.3389/fimmu.2014.00047

Lu J, Chang P, Richardson JA, Gan L, Weiler H, Olson EN (2000) The basic helix-loop-helix transcription factor capsulin controls spleen organogenesis. Proc Natl Acad Sci U S A 97 (17):9525–9530

Lyons I, Parsons LM, Hartley L, Li R, Andrews JE, Robb L, Harvey RP (1995) Myogenic and morphogenetic defects in the heart tubes of murine embryos lacking the homeo box gene Nkx2-5. Genes Dev 9(13):1654–1666

Mackay F, Majeau GR, Lawton P, Hochman PS, Browning JL (1997) Lymphotoxin but not tumor necrosis factor functions to maintain splenic architecture and humoral responsiveness in adult mice. Eur J Immunol 27(8):2033–2042. doi:10.1002/eji.1830270830

Mahlaoui N, Minard-Colin V, Picard C, Bolze A, Ku CL, Tournilhac O, Gilbert-Dussardier B, Pautard B, Durand P, Devictor D, Lachassinne E, Guillois B, Morin M, Gouraud F, Valensi F, Fischer A, Puel A, Abel L, Bonnet D, Casanova JL (2011) Isolated congenital asplenia: a French nationwide retrospective survey of 20 cases. J Pediatr 158(1):142–148. doi:10.1016/j.jpeds.2010.07.027

Mebius RE, Kraal G (2005) Structure and function of the spleen. Nat Rev Immunol 5(8):606–616. doi:10.1038/nri1669

Miller JP, Yeh N, Vidal A, Koff A (2007) Interweaving the cell cycle machinery with cell differentiation. Cell Cycle 6(23):2932–2938

Mueller SN, Germain RN (2009) Stromal cell contributions to the homeostasis and functionality of the immune system. Nat Rev Immunol 9(9):618–629. doi:10.1038/nri2588

Ngo VN, Cornall RJ, Cyster JG (2001) Splenic T zone development is B cell dependent. J Exp Med 194(11):1649–1660

Patterson KD, Drysdale TA, Krieg PA (2000) Embryonic origins of spleen asymmetry. Development 127(1):167–175

Prall OW, Menon MK, Solloway MJ, Watanabe Y, Zaffran S, Bajolle F, Biben C, McBride JJ, Robertson BR, Chaulet H, Stennard FA, Wise N, Schaft D, Wolstein O, Furtado MB, Shiratori H, Chien KR, Hamada H, Black BL, Saga Y, Robertson EJ, Buckingham ME, Harvey RP (2007) An Nkx2-5/Bmp2/Smad1 negative feedback loop controls heart progenitor specification and proliferation. Cell 128(5):947–959. doi:10.1016/j.cell.2007.01.042

Quaggin SE, Schwartz L, Cui S, Igarashi P, Deimling J, Post M, Rossant J (1999) The basic-helix-loop-helix protein pod1 is critically important for kidney and lung organogenesis. Development 126(24):5771–5783

Roberts CW, Sonder AM, Lumsden A, Korsmeyer SJ (1995) Development expression of Hox11 and specification of splenic cell fate. Am J Pathol 146(5):1089–1101

Scandella E, Bolinger B, Lattmann E, Miller S, Favre S, Littman DR, Finke D, Luther SA, Junt T, Ludewig B (2008) Restoration of lymphoid organ integrity through the interaction of lymphoid tissue-inducer cells with stroma of the T cell zone. Nat Immunol 9(6):667–675. doi:10.1038/ni.1605

Selleri L, Depew MJ, Jacobs Y, Chanda SK, Tsang KY, Cheah KS, Rubenstein JL, O'Gorman S, Cleary ML (2001) Requirement for Pbx1 in skeletal patterning and programming chondrocyte proliferation and differentiation. Development 128(18):3543–3557

Suematsu S, Watanabe T (2004) Generation of a synthetic lymphoid tissue-like organoid in mice. Nat Biotechnol 22(12):1539–1545. doi:10.1038/nbt1039

Tammela T, Saaristo A, Holopainen T, Lyytikka J, Kotronen A, Pitkonen M, Abo-Ramadan U, Yla-Herttuala S, Petrova TV, Alitalo K (2007) Therapeutic differentiation and maturation of lymphatic vessels after lymph node dissection and transplantation. Nat Med 13 (12):1458–1466. doi:10.1038/nm1689

Tan JK, Watanabe T (2014) Murine spleen tissue regeneration from neonatal spleen capsule requires lymphotoxin priming of stromal cells. J Immunol 193(3):1194–1203. doi:10.4049/jimmunol.1302115

Tribioli C, Lufkin T (1999) The murine Bapx1 homeobox gene plays a critical role in embryonic development of the axial skeleton and spleen. Development 126(24):5699–5711

Vondenhoff MF, Greuter M, Goverse G, Elewaut D, Dewint P, Ware CF, Hoorweg K, Kraal G, Mebius RE (2009) LTbetaR signaling induces cytokine expression and up-regulates lymphangiogenic factors in lymph node anlagen. J Immunol 182(9):5439–5445. doi:10.4049/jimmunol.0801165

Withers DR, Kim MY, Bekiaris V, Rossi SW, Jenkinson WE, Gaspal F, McConnell F, Caamano JH, Anderson G, Lane PJ (2007) The role of lymphoid tissue inducer cells in splenic white pulp development. Eur J Immunol 37(11):3240–3245. doi:10.1002/eji.200737541

Yokota Y, Mansouri A, Mori S, Sugawara S, Adachi S, Nishikawa S, Gruss P (1999) Development of peripheral lymphoid organs and natural killer cells depends on the helix-loop-helix inhibitor Id2. Nature 397(6721):702–706. doi:10.1038/17812

Chapter 6
Controlling Lymphoid Development

Sergio A. Lira and Glaucia C. Furtado

Abstract Formation of lymphoid tissue involves complex cellular and molecular interactions. In this chapter we will review our recent work on the role of chemokines and TNFα in lymphoid organogenesis.

Introduction

Lymphoid organs are critical for generation of adaptive immune response. Secondary lymphoid organs (SLO) are formed at predefined areas during embryogenesis whereas tertiary lymphoid organs (TLO) are formed after birth in tissues with ongoing inflammatory processes (Randall et al. 2008; Neyt et al. 2012). Both secondary and tertiary lymphoid organs have lymphocytes that are topologically segregated, and diverse sets of myeloid and stromal cells. In addition, they have specialized vasculature such as high endothelial venules (HEV), and a lymphatic network. In this chapter we will review our recent work on the role of chemokines and TNFα in lymphoid organogenesis.

Secondary Lymphoid Organs (SLO)

Lymphocytes are born and mature in the bone marrow and thymus (primary lymphoid organs). Once their maturation is complete they reach the circulation and migrate continuously through SLO (lymph nodes, spleen, tonsils, Peyer's patches, and mucosal associated lymphoid tissue (MALT)). SLO are critical for the immune response because they are the sites where the lymphocytes are activated.

S.A. Lira (✉) • G.C. Furtado
Immunology Institute, Icahn School of Medicine at Mount Sinai, New York, NY 10029, USA
e-mail: sergio.lira@mssm.edu; glaucia.furtado@mssm.edu

T. Watanabe, Y. Takahama (eds.), *Synthetic Immunology*,
DOI 10.1007/978-4-431-56027-2_6

Mechanisms involved in the development of SLO. The two major cell types involved in lymph node organogenesis are the hematopoietic lymphoid tissue inducer (LTi) cells and non-hematopoietic lymphoid tissue stromal "organizer cells" (LTo) (Randall et al. 2008). The current model for development of lymphoid organs suggests that LTi cells originate in the fetal liver from common lymphoid progenitors and that they migrate to the sites where the lymph nodes are formed (lymph node anlagen) (Randall et al. 2008; Roozendaal and Mebius 2011) (Fig. 6.1). Clustering of LTi and LTo cells is an essential step in lymph node development (Kim et al. 2000). Animals that are deficient in the nuclear retinoid orphan receptor (ROR)γ, encoded by the *Rorc* gene, or the inhibitor of DNA binding 2, dominant negative helix-loop-helix protein Id2, lack LTi cells and therefore fail to form lymph nodes and Peyer's patches (Sun et al. 2000; Eberl and Littman 2004; Yokota et al. 1999). At lymphoid anlagen, binding of the TNFα family ligand Receptor Activator of NF-kB (RANKL) to its receptor RANK induces the differentiation and survival of LTi cells and trigger expression of lymphotoxin-alpha-1-beta-2 (LTα1β2) on their surface (Kong et al. 1999; Dougall et al. 1999; Kim et al. 2000; De Togni et al. 1994; Vondenhoff et al. 2009). A key step in the development of LNs is the engagement of LTα1β2 expressed by LTi cells to its receptor lymphotoxin-beta receptor (LTβR) on mesenchymal organizer cells (Futterer et al. 1998; Crowe et al. 1994). This interaction promotes upregulation of intracellular adhesion molecule (ICAM-1), vascular cell adhesion molecule (VCAM-1) and mucosal addressin cell adhesion molecule (MAdCAM-1) on the surface of LTo cells (Cupedo et al. 2004; Okuda et al. 2007) and the expression of the chemokines CCL19, CCL21, and CXCL13 (Roozendaal and Mebius 2011). CCL19, CCL21, and CXCL13, play an important role in the formation of secondary lymphoid organs and Payer's patches, via interaction with

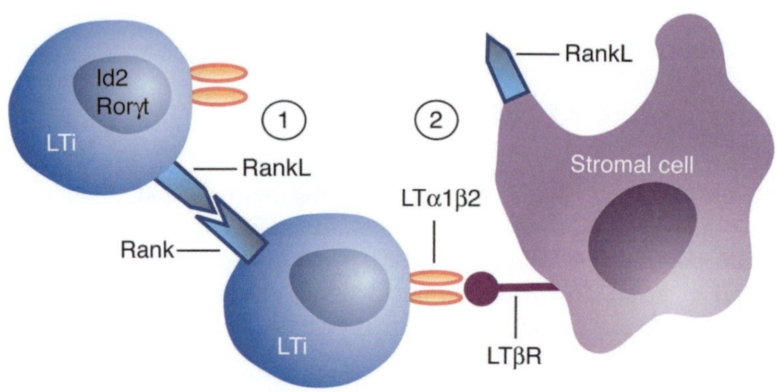

Fig. 6.1 Model of SLO formation. LTi precursor cells migrate from the fetal liver to the lymph node anlagen. (**1**) LTi RANK and RANK ligand interaction induces the upregulation of LTα1β2 receptor on the surface of LTi cells. (**2**) LTα1β2 receptor binds to stromal cell LTβR. This interaction induces the differentiation of stromal cells and the induction of the homeostatic chemokines CXCL13 and CCL19/CCL21. These chemokines attract additional LTi cells to the LN anlagen and T and B cells to the LN after birth

their receptors CCR7 and CXCR5 (Okada et al. 2002; Warnock et al. 2000; Forster et al. 1999; Drayton et al. 2006; Mebius 2003). CXCL13 appears to contribute to the formation of SLO in two ways: mediating the influx of mature lymphocytes into the newly forming follicles and regulating the recruitment of CXCR5^{+} LTi cells to the anlagen (Mebius et al. 1997).

Animals genetically deficient in LTα and LTβR do not form lymph nodes or Peyer's patches (De Togni et al. 1994; Banks et al. 1995; Futterer et al. 1998). Furthermore, genetic deletion of molecules in the LTβR signaling pathway (NF-kappa B non canonical pathway) such as NF-kappa B-inducible kinase (NIK) (Miyawaki et al. 1994) and RelB (Weih and Caamano 2003) precludes LN formation. While the role of LTα1β2/LTβR is firmly established in the process of lymphoid organogenesis, the role of other members of the TNFα superfamily is unclear. TNFα or TNF-R1-deficient mice have all lymph nodes, including mLN, but they fail to form B cell follicles. These results suggest that TNFα activity in lymphoid organogenesis may be secondary to other TNFα members such as LT. However, simultaneous deficiency of TNFR1 and RelA blocks the development of all lymph nodes, despite the presence of a normal complement of LTα1β2^{+} LTi cells (Alcamo et al. 2002). Female mice injected in utero with LTβR-Ig fusion protein retain cervical and mesenteric lymph nodes (mLN) but fail to form other lymph nodes (Rennert et al. 1996, 1997). However, simultaneous treatment LTβR-Ig fusion protein and anti-TNFR1 antibody, or LTβR-Ig plus anti-TNFα antibodies, prevents development of all lymph nodes (Rennert et al. 1998), which suggests that TNFα has a role in mLN organogenesis.

Tertiary Lymphoid Organs (TLO)

Tertiary lymphoid organs (TLO) are lymphoid structures that resemble SLO, but differ in that they do not show a capsule and form in sites of chronic inflammation. TLOs are frequently found in tissues affected by autoimmunity, including thyroid, pancreas, liver, salivary glands, thymus and stomach (for review see (Neyt et al. 2012)). The recent detection of these structures within neoplasms (Goc et al. 2013) suggests that TLO may also play a role in tumor biology.

Mechanisms involved in development of TLO. Several reports in the literature link chemokines to development of TLO. CCL21, a potent chemoattractant for lymphocytes and dendritic cells is expressed in the thyroid of patients with autoimmune thyroid disease (AITD). One of the hallmarks of AITD, especially Hashimoto's thyroiditis, is the presence of large TLO (Fig. 6.2a–d). Studies done by our laboratory using transgenic mice indicate that expression of CCL21 in the thyroid is sufficient to promote development of TLO (Fig. 6.2e–h). Within these TLO we observed well-segregated B and T cell areas, and specialized vasculature (PNAd^{+} high-endothelial venules and LYVE-1^{+} lymphatics) (Martin et al. 2004). Although high endothelial venules expressing peripheral lymph node addressin were frequently observed in the thyroid infiltrates, lymphocyte recruitment was

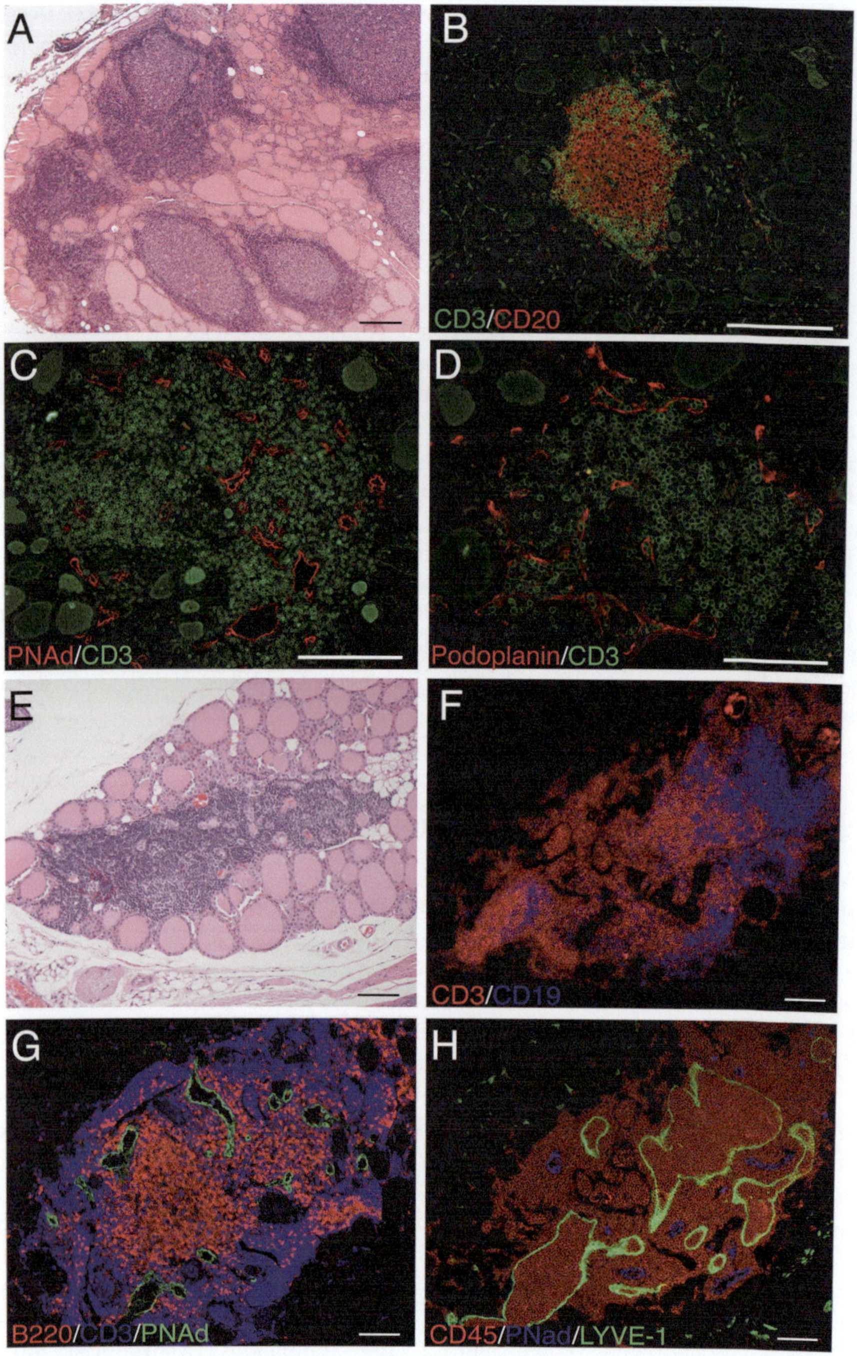

Fig. 6.2 Expression of CCL21 in the thyroid of transgenic mice leads to formation of TLO similar to those found in human AITD. (**a–d**) Sections from the thyroid of a patient with Hashimoto's thyroiditis (**a**) H&E stained section showing the presence of mononuclear cell aggregates. (**b**) The infiltrates contain T (green) and B (red) cells that are topologically segregated.

independent of L-selectin or LTα but required CCR7 expression. Mechanistically the formation of CCL21-induced TLO in the thyroid did not depend on the presence of lymphoid tissue-inducer cells expressing the *Id2* gene, essential for generation of $CD3^-CD4^+$ LTi cells and development of secondary lymphoid organs (Marinkovic et al. 2006). Rather, the formation of the TLO was dependent on mature $CD3^+CD4^+$ T cells. The initial stages of TLO formation involved interaction of $CD3^+CD4^+$ T cells with DCs, the appearance of peripheral-node addressin–positive (PNAd+) vessels, and production of chemokines that recruit lymphocytes and DCs and lymphangiogenesis (Fig. 6.3). Genetic deletion of LTβR receptor or LTα abrogated development of lymphatic vessels in the inflamed areas in the thyroid but did not affect development of neighboring lymphatics (Furtado et al. 2007). The development of lymphatic vessels within the TLO was dependent on the expression of lymphotoxin ligands by host cells, but not by the transferred CD4(+) T cells. Ablation of host DCs, but not natural killer (NK) cells, reduced the formation of new lymphatic vessels in the thyroid (Muniz et al. 2011). These results suggest that the formation of CCL21-induced TLO involves not only recruitment of lymphocytes and DC, but also specific vascular changes that are consequent to the influx of these cells into the tissue.

TLO are also frequently detected in non-obese diabetic (NOD) mice, in particular in salivary glands and in the pancreas. CCL21 is expressed in para-ductal areas and within TLO present in the pancreas of pre-diabetic NOD mice (Martin et al. 2008) (Fig. 6.4). Transgenic mice that express CCL21 in the pancreatic islets also develop TLO rich in T and B cells that are topologically segregated (Chen et al. 2002b; Fan et al. 2000; Luther et al. 2002). The infiltrates also contained $CD11c^+$ dendritic DC and very small numbers of macrophages ($F4/80^+$ cells) (Chen et al. 2002b). Pancreatic TLOs observed in these animals resembled those observed in NOD mice in the pre-diabetic stage. Similar to other TLO, mucosal addressin (MAdCAM-1) and peripheral lymph node addressin (PNAd), adhesion molecules specifically expressed in secondary lymphoid structures, were also observed in these infiltrates, along with ER-TR7 positive cells characteristic of lymph node stromal reticulum (Chen et al. 2002b). Thus pancreatic expression of CCL21 is sufficient to induce the formation of TLO in the pancreas. However, the ability of CCL21 to promote recruitment of lymphocytes and the formation of TLO appears to be tissue-specific, because overexpression of CCL21 in the skin (Chen et al. 2002b) or brain (Chen et al. 2002a) does not promote lymphocyte recruitment, suggesting that there may be tissue-specific requirements for its activity. This

Fig. 6.2 (continued) Within the lymphoid aggregates (green) there are (**c**) $PNAd^+HEV$ (red) and (**d**) $podoplanin^+$ (red) lymphatic vessels, similar to what is seen in lymph nodes. (**e**–**h**) Expression of CCL21 in the thyroid of transgenic mice (TGCCL21 mice) induces development of TLO. (**e**) H&E stained section of a TGCCL21 thyroid showing mononuclear cell aggregates. (**f**) Topologically segregated T (red) and B (blue) cells are found within the thyroid. (**g**) $PNAd^+HEV$ vessels (green) and (**h**) $LYVE\text{-}1^+$ lymphatic vessels (green) are also present within the infiltrates. Scale bars = 250 μm

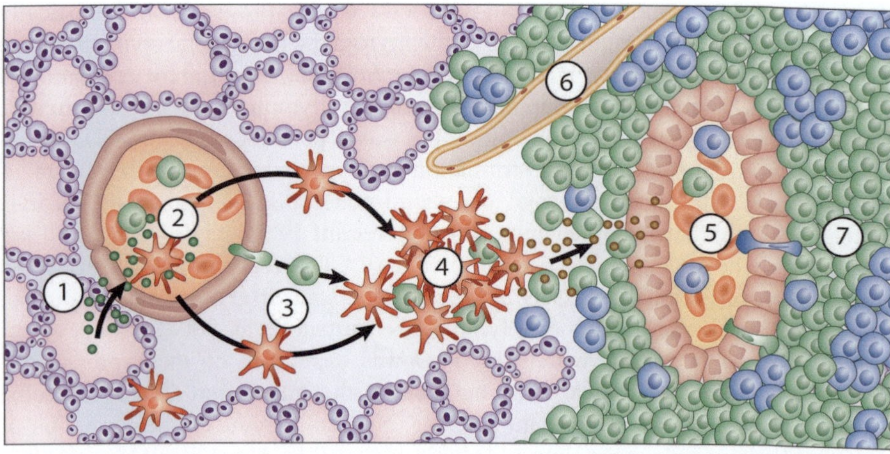

Fig. 6.3 A model for the formation of TLO in the thyroid of TGCCL21 mice. Expression of CCL21 by thyrocytes (**1**) promotes recruitment of $CD3^{+}CD4^{+}$ T cells from venules (**2**). Entry of T cells into the tissue (**3**) induces recruitment of dendritic cells (DC), which cluster with T cells (**4**). Chemokines, cytokines and other factors contribute to development of specialized vascular structures such as HEV (**5**) and lymphatics (**6**) and to the recruitment and topological arrangement of T cells and B cells (**7**) (Model based on findings described in Marinkovic et al. (2006))

tissue-specificity may relate to the nature of the vasculature supplying these tissues. The vascular bed of the endocrine pancreas, like that of other endocrine glands such as the thyroid, is unique, and may be permissive to CCL21's activity in inducing lymphocyte migration.

Another chemokine involved in the formation of TLO is CXCL13, a chemokine that promotes chemoattraction of B cells in vitro and in vivo (Martin et al. 2006). Studies done by our group have shown that CXCL13, besides being expressed constitutively in lymphoid tissues, is also expressed during inflammatory conditions in organs such as the pancreas (Martin et al. 2008) and the gut (Shang et al. 2009). In the intestine of adult mice two general types of lymphoid aggregates can be found: isolated lymphoid follicles (ILF) and TLO. ILFs are genetically programmed clusters of B cells present at the base of the villi, that require $ROR\gamma t^{+}$ LTi cells and LTβR signaling for their formation (Eberl and Littman 2004; Eberl et al. 2004; Tsuji et al. 2008; Lorenz et al. 2003). TLO in the gut are composed by large clusters of $B220^{+}$ cells that contain $CD3^{+}$ lymphocytes, and are formed in response to infection or inflammation (Lochner et al. 2011; Picarella et al. 1992). Transgenic expression of CXCL13 in the intestine promoted a marked increase in the number of B and NK cells in the lamina propria and an increase in the size and number of lymphoid follicles in the small intestine (Marchesi et al. 2009). This process was associated with the presence of cells expressing markers found in LTi, such as LTα, LTβ, and TNF-related activation-induced cytokine (TRANCE). Such LTi-like cells produced IL-22, a cytokine implicated in epithelial repair; and expressed the IL-23 receptor, a key regulator of IL-22 production. Whether these cells had a direct role in promoting development of TLO is unknown at present.

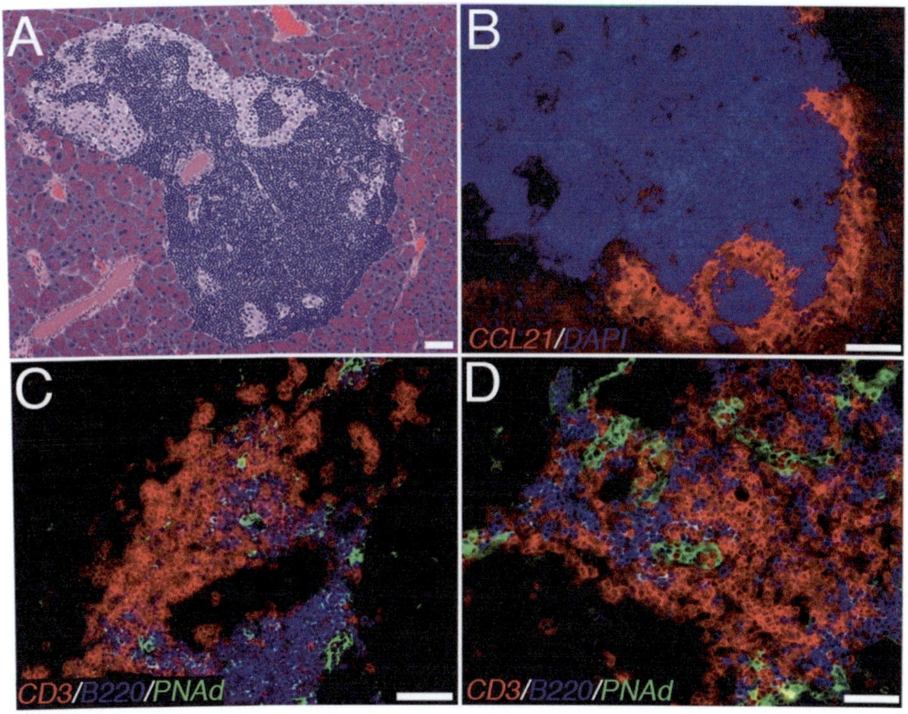

Fig. 6.4 Expression of CCL21 in pancreas induces development of TLO. (**a**) H&E stained paraffin section of RIPCCL21 pancreas showing cellular infiltrates in the islets. (**b**) Expression of CCL21 in the islets. Note the presence of a large infiltrate (DAPI, *blue*). (**c**) Cell infiltrates in the pancreatic islets of RIPCCL21 mice were analyzed by immunostaining with labeled anti-CD3 (*red*) anti-B220 (*blue*). PNAd staining (*green*) labels HEV. Topological segregation of T and B cells and presence of HEV were also seen in the islets of NOD mice at 10 weeks of age (**d**). Scale bars = 250 μm

Role of TNFα in Lymphoid Organogenesis

As discussed above, the role of TNFα in lymphoid neogenesis is incompletely understood. To define the role of LTi cells and TNFα in the formation of lymphoid aggregates in the intestine we examined the presence of these structures in the ileum of $TNF^{\Delta ARE/+}$ mice. These animals express increased levels of TNFα under basal conditions, due to a deletion in the 3' region of the *Tnfa* gene that results in higher stability of its mRNA and, consequently, increased levels of TNFα protein (Kontoyiannis et al. 2002). $TNF^{\Delta ARE/+}$ mice develop marked inflammatory infiltrates in the joints and in the intestine. The inflammatory infiltrates in the intestine are located in the terminal ileum and are composed of neutrophils, macrophages, and T cells that are distributed throughout the submucosa and muscular layers and sometimes reach the serosa. Large mononuclear aggregates rich in B cells, or TLO, are also found in the terminal ileum of the $TNF^{\Delta ARE/+}$ mice (Viejo-Borbolla

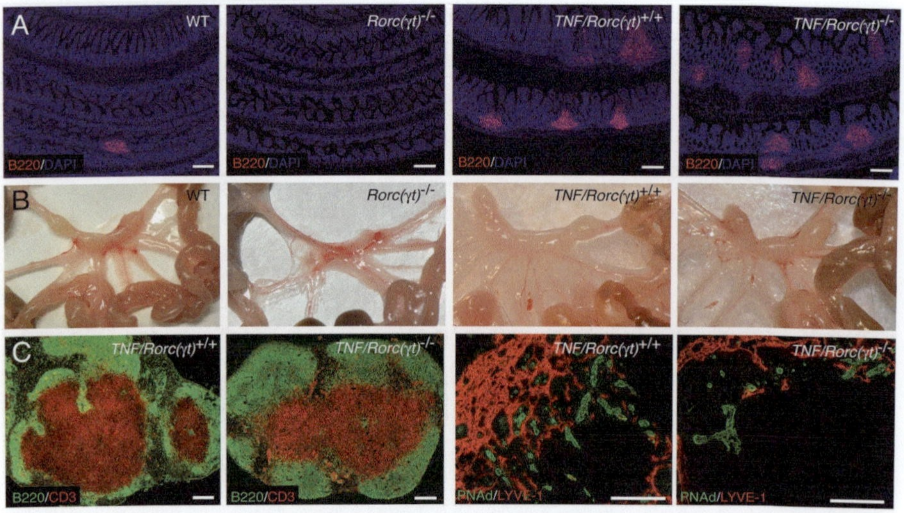

Fig. 6.5 TLO and SLO are formed in the ileum *TNF/Rorc(γt)*$^{-/-}$ mice. (**a**) Ileum sections of indicated mice stained with anti-B220 antibody to visualize B cell aggregates and DAPI for nuclear staining. Small B cell clusters were found in the ileum of WT but were absent in the ileum of *Rorc(γt)*$^{-/-}$ mice. Overexpression of TNF induced the formation of large B cell clusters with few T cells in the ileum of *TNF/Rorc(γt)*$^{+/+}$ and *TNF/Rorc(γt)*$^{-/-}$ mice. (**b**) Increased expression of TNFa induces development SLO in the absence of RORγt^{+} LTi cells. Photograph of the mesentery of WT, *Rorc(γt)*$^{-/-}$, *TNF/Rorc(γt)*$^{+/+}$, and *TNF/Rorc(γt)*$^{-/-}$ mice. (**c**) Lymph nodes from *TNF/Rorc(γt)*$^{+/+}$ and *TNF/Rorc(γt)*$^{-/-}$ mice at 6 weeks of age were analyzed by immunostaining. Note segregation of T and B cell areas, presence of PNAd^{+} HEV and lymphatic vessels, in the mLN lymph nodes of *TNF/Rorc(γt)*$^{-/-}$ mice. These features were similar to the ones observed in the mLn of *TNF/Rorc(γt)*$^{+/+}$ mice. Scale bars = 250 μm (Adapted from "TNFa-dependent development of lymphoid tissue in the absence of RORgt^{+} lymphoid tissue inducer cells", by Furtado et al. (2014))

et al. 2010). To determine whether the formation of these aggregates is dependent on RORγt^{+}LTi cells we crossed *Rorc(γt)*$^{-/-}$ mice with *TNF*$^{\Delta ARE/+}$ mice to generate *TNF/Rorc(γt)*$^{-/-}$ mice (Furtado et al. 2014). Histological analysis of the terminal ileum of age-matched wild type (WT), *Rorc(γt)*$^{-/-}$, *TNF/Rorc(γt)*$^{+/+}$ *and* *TNF/Rorc(γt)*$^{-/-}$ mice at 16–20 weeks of age showed that *TNF/Rorc(γt)*$^{+/+}$ and *TNF/Rorc(γt)*$^{-/-}$ mice, but not WT or *Rorc(γt)*$^{-/-}$ mice, had marked submucosal inflammation, vilus blunting, patchy transmural inflammation, and lymphoid aggregates. The lymphoid aggregates in *TNF/Rorc(γt)*$^{+/+}$ and *TNF/Rorc(γt)*$^{-/-}$ mice contained large clusters of B220^{+} B cells and few CD3^{+} T cells, which were absent in *Rorc(γt)*$^{-/-}$ mice. These results indicated that TLO can be formed in the ileum in the absence of RORγt^{+} LTi cells.

During our phenotypic analyses of the *TNF/Rorc(γt)*$^{-/-}$ mice we observed that 100 % of the *TNF/Rorc(γt)*$^{-/-}$ mice developed mesenteric LN (mLN) that were grossly indistinguishable from those found in WT mice. The phenomenon was not restricted to the mLN; axillary, cervical, brachial, inguinal, para-aortic, and peripancreatic LN were also present at varying frequencies. Mediastinal and

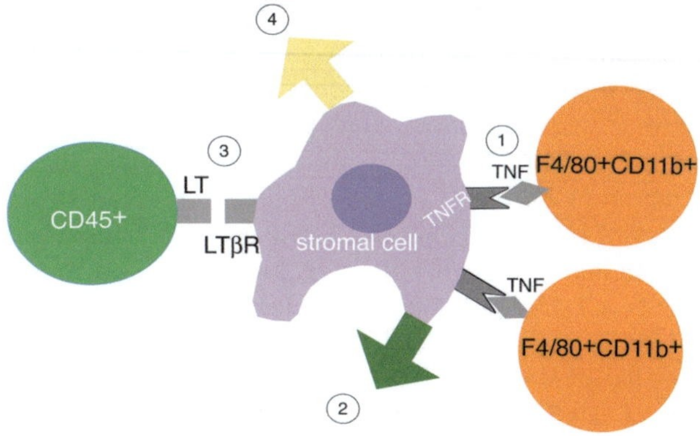

Fig. 6.6 A model of LN development in *TNF/Rorc(γt)*$^{-/-}$ mice. Increased expression of TNFa by F4/80^{+}CD11b^{+} cells is sufficient to maintain lymph node stromal cells up to early postnatal life (**1**). Stromal cells produce chemokines that are important for cell recruitment (**2**). After birth, lymphoid cells and myeloid cells expressing LTa are recruited into the anlagen and trigger LTbR signaling (**3**), which promotes expression of additional chemokines, cell adhesion molecules, and other unidentified factors (**4**). These events are critical for further recruitment of lymphoid cells and proper differentiation of HEVs and lymphatics. Model based on findings described in Furtado et al. (2014)

popliteal LN, as well as Peyer's patches, were not observed in these animals. LNs of WT and *TNF/Rorc(γt)*$^{-/-}$ mice had segregated T and B cell areas, PNAd^{+} high endothelial venules, an extensive lymphatic network, ER-TR7^{+} lymph node stroma and CD35^{bright} follicular dendritic cells. When immunized with OVA, these animals had serum levels of OVA-specific IgA and IgG similar to those observed in *TNF/Rorc(γt)*$^{+/+}$ mice (Furtado et al. 2014). These results indicated that the SLO formed in the *TNF/Rorc(γt)*$^{-/-}$ mice were anatomically and functionally indistinguishable from those found in *TNF/Rorc(γt)*$^{+/+}$ mice. These findings were unexpected because, LTis, which are absent in *TNF/Rorc(γt)*$^{-/-}$ mice, were thought to be absolutely required for development of SLO.

The presence of mLN in the *TNF/Rorc(γt)*$^{-/-}$ mice suggested that a RORγt-independent cell type could play a role in the formation of SLO. Analysis of the LN cellularity at P1 showed that F4/80^{+}, NK1.1^{+} and CD11c^{+} cells were the most abundant CD45^{+} leukocytes present in the mLN anlagen of *Rorc(γt)*$^{-/-}$ and *TNF/Rorc(γt)*$^{-/-}$ mice. Further studies showed that TNFα was physiologically expressed by macrophages in the anlagen during development and that its levels were elevated in the mLN of *TNF/Rorc(γt)*$^{-/-}$ mice compared to *Rorc(γt)*$^{-/-}$ mice. Importantly, the source of increased TNFα levels in the anlagen of the *TNF/Rorc(γt)*$^{-/-}$ mice were macrophages.

To determine how TNFα expression by myeloid cells could contribute to lymph node organogenesis, we compared the transcriptomes of the mLN anlagen of *Rorc(γt)*$^{-/-}$ and *TNF/Rorc(γt)*$^{-/-}$ mice at postnatal day 1 (P1). Expression of several genes involved in lymphoid organogenesis such as *Cxcl13*, *Lymphotoxin beta*, *Relb*, *Ccl19* and *Madcam-1*, was increased in the mLN of *TNF/Rorc(γt)*$^{-/-}$ mice suggesting that expression of TNFα contributed to the generation of a lymphoid organogenic program. In addition to these changes we detected changes in the stroma, a key step in lymphoid organogenesis (Randall et al. 2008). Activated (ICAM-1^{hi}VCAM-1^{hi}) stromal cells were present at a significantly higher frequency in the mLN of *TNF/Rorc(γt)*$^{-/-}$ mice at E18.5 onwards when compared with the same region of *Rorc(γt)*$^{-/-}$ mice (Benezech et al. 2010) . One day after birth the frequency of ICAM-1^{hi}VCAM-1^{hi} cells in mLN stroma of *TNF/Rorc(γt)* $^{-/-}$ was higher than that of *Rorc(γt)*$^{-/-}$ mice, but comparable to that of *WT* mice. Sorted stromal cells (CD45-) from the mLN anlagen of *TNF/Rorc(γt)*$^{-/-}$ mice at P1 expressed increased levels of Cxcl13, Ccl19 and Ccl21 mRNA when compared to sorted stromal cells of *Rorc(γt)*$^{-/-}$ mice. Together these results indicated that increased levels of TNFα were sufficient to induce lymph node stromal cell maturation in the absence of LTi cells.

The high expression levels of LTβR ligands in the mLN of *TNF/Rorc(γt)*$^{-/-}$ mice after birth suggested a role for Lymphotoxin-LTβR signaling in the development of LN in this model. To determine if this was indeed the case, we crossed *TNF* $^{\Delta ARE/+}$ mice with *LTβR*-deficient animals (referred to as *TNF/LTβR*$^{-/-}$ mice). With exception of mLN, no lymph nodes, Peyer's Patches and TLO were found in any of the *TNF/LTβR*$^{-/-}$ mice animals examined. The mLN of *TNF/LTβR*$^{-/-}$ mice were markedly abnormal as shown by the absence of B cell follicles and T cell areas and the impaired recruitment of lymphocytes to these organs. In addition, HEVs were absent and the number of lymphatic vessels was reduced. Finally, these abnormal mLN lacked CD35^{bright} follicular dendritic cells, and had an aberrant ER-TR7^{+} stroma. Thus, TNFα overexpression was not able to rescue LN development in mice lacking LTβR signaling.

Finally, we investigated if TNFα would bypass the requirement for *Id2* in lymph node organogenesis. Id2-deficient mice lack LTi cells (Yokota et al. 1999), NK cells (Yokota et al. 1999) and fetal CD11b^{+} myeloid cells (Eberl and Littman 2004) in the lymph node anlagen, and do not develop SLO. None of the *TNF/Id2*$^{-/-}$ mice examined at birth had mLN. Of interest, we found that F4/80^{+} myeloid cells were present in the mLN anlagen of WT and *TNF/Id2*$^{+/-}$ mice but absent in *TNF/Id2*$^{-/-}$ mice, and in *Id2*$^{-/-}$ mice, in agreement with previous reports (Eberl and Littman 2004). Myeloid cell migration to the mLN anlagen of *TNF/Id2*$^{-/-}$ mice was impaired and, strikingly, no LNs were formed in these mice. Thus, TNFα does not bypass the requirement for *Id2* in lymphoid organogenesis.

Conclusions

Several studies support the concept that SLO and TLO development share common mechanisms. However, evidence first derived from analysis of CCL21-driven transgenic models (Marinkovic et al. 2006) and other models (Rangel-Moreno et al. 2011), suggested that TLO development does not have a requirement for LTi cells which are required for SLO development under homeostatic conditions.

Our studies and those of Eberl's group (Lochner et al. 2011) demonstrate that inflammatory stimuli promote development of TLO in the ileum and colon in the absence of RORγt$^+$ LTi cells, and implicate TNFα as an important factor in their generation, since its expression is elevated in the ileum of $TNF^{\Delta ARE/+}$ mice and in the colon of DSS-treated animals (Viejo-Borbolla et al. 2010; Kojouharoff et al. 1997). They also suggest that formation of TLOs in the intestine is independent of Th17$^+$ cells and other *Rorc(γt)*-dependent innate lymphoid cells. Surprisingly, our recent work shows that SLO development can occur in the absence of *Rorc*-dependent LTi cells, provided there is dysregulated expression of TNFα.

How could TNFα promote organogenesis in the absence of LTi cells?

We suggest that increased levels of TNFα functionally compensates for the lack of LTβR signaling during embryogenesis and contributes to the maintenance of a functional anlagen. This hypothesis is supported by our observations that stromal cells in the anlagen of $TNF/Rorc(\gamma t)^{-/-}$, but not in the $Rorc(\gamma t)^{-/-}$ mice, are activated. $TNF/Rorc(\gamma t)^{-/-}$ stromal cells express higher levels of ICAM-1 and VCAM-1 during embryogenesis and immediately after birth. Furthermore, cells in the $TNF/Rorc(\gamma t)^{-/-}$ anlagen express increased levels of the TNFα-inducible chemokines CCL2 and CXCL10. These chemokines, acting in concert with TNFα, could promote recruitment of additional hematopoietic cells. At birth, influx of hematopoietic cells could further contribute to organogenesis. Interestingly, we noted that the expression of LT ligands increased after birth. This could reflect either increased expression of LT ligands by resident non-LTi cells or reflect increased influx of hematopoietic cells that express LT ligands. We favor the second hypothesis because we have observed increased influx of hematopoietic cells in the $TNF/Rorc(\gamma t)^{-/-}$ anlagen during the first 5 days of life.

The increased expression of LT ligands is absolutely critical for normal lymph node development as shown by the analysis of the TNF/LTβR mutants. In the newborn $TNF/Rorc(\gamma t)^{-/-}$ anlagen, LT ligands could potentially synergize with TNFa to activate the transcription of several molecules related to lymph node organogenesis, macrophage function, and inflammation. A recent report has shown the synergistic effect of TNFα signaling together with the alternative NF-kB pathway to drive high expression levels of Ccl21, Cxcl13, Vcam-1, Icam-1 and Madcam-1 in spleens (Lovas et al. 2012). Of note, we have detected expression of Cxcl13, Ccl19 and Ccl21, by stromal cells located in the mLN of $TNF/Rorc(\gamma t)^{-/-}$ mice at P1. These chemokines induce lymphoid organogenesis when expressed in vivo (Luther et al. 2000; Martin et al. 2006; Marchesi et al. 2009), and their expression by stromal cells in the anlagen could account for the formation of LN in $TNF/Rorc(\gamma t)^{-/-}$mice.

Our results suggest that myeloid cells may contribute to development of SLO, working in part as LTi-like cells. Myeloid cells are present in the anlagen of WT mLN as early as E15.5 and produce TNF. The generation of the CD11b$^+$ cells does not require RORgt because these myeloid cells are still present in the LN anlagen of $Rorc(\gamma t)^{-/-}$ mice (Eberl and Littman 2004). The increased number of myeloid cells that express stable *tnfa* mRNA contribute to high levels of this protein in the lymph node anlagen in TNF/$Rorc(\gamma t)^{-/-}$ mice. Consistent with a role for myeloid cells in lymph node development in the TNF/*Rorc(γt)*$^{-/-}$ mouse model is the observation that absence of these cells (and thus the source of TNFα necessary to activate the stromal cells) in lymph node anlagen of *TNF/Id2*$^{-/-}$ mice resulted in the failure to rescue the formation of these organs.

TNFα receptor ligation activates the NF-κB classical pathway, which involves the IκB kinase and results in the activation of RelA. LTβR ligation activates both the NF-κB classical and alternative pathways (Weih and Caamano 2003). The alternative NF-κB pathway is mediated by the NF-κB-inducing kinase (NIK) and results in the activation of NF-kB2/Relb. Because animals genetically deficient in *LtbR, Nik*, and *RelB* do not form lymph nodes (Futterer et al. 1998; Miyawaki et al. 1994; Weih and Caamano 2003) it was concluded that the alternative pathway is critically important for the generation of SLO (Weih and Caamano 2003). However, simultaneous deletion of TNFR1 and RelA precludes formation of all LN and PP in double knockout mice due to a stromal cell defect, even in the presence of LTi cells expressing normal levels of Lymphotoxin (Alcamo et al. 2002), which suggests that the canonical NF-kB pathway is physiologically important for normal development. LTβR signaling has a profound effect in the generation of most LN and intestinal TLO, a role that cannot be bypassed even in the presence of increased levels of TNFα. While increased TNFα cannot compensate for the lack of LTβR in terms of TLO and SLO development, it can partially compensate for lack of LTβR-signaling in the development of mLN. Taken together, our studies highlight a significant cross-talk between these receptor systems for the development and function of lymphoid structures.

In summary: Our studies reveal an important role for the chemokines CCL21 and CXCL13 in the development of TLO. When expressed ectopically, these chemokines are sufficient to induce generation of TLO in many, but not all tissues. Our results also suggest an important role for TNFα in inducing development of TLO and indicate that increased TNFα expression can promote development of TLO and SLO, in the absence of canonical Rorc + LTis. Thus, our results show that lymphoid tissue organogenesis can occur in the absence of Rorγt^{+} LTi cells and suggest that interactions between TNFα-expressing myeloid cells and stromal cells have an important role in this process. Identifying the precise nature of the stromal cell (fibroblast, endothelial cell, etc.) that serves as "the organizer" in this setting is critical to understand how TNFα affects lymphoid neogenesis.

References

Alcamo E, Hacohen N, Schulte LC, Rennert PD, Hynes RO, Baltimore D (2002) Requirement for the NF-kappaB family member RelA in the development of secondary lymphoid organs. J Exp Med 195(2):233–244

Banks TA, Rouse BT, Kerley MK, Blair PJ, Godfrey VL, Kuklin NA, Bouley DM, Thomas J, Kanangat S, Mucenski ML (1995) Lymphotoxin-alpha-deficient mice. Effects on secondary lymphoid organ development and humoral immune responsiveness. J Immunol 155 (4):1685–1693

Benezech C, White A, Mader E, Serre K, Parnell S, Pfeffer K, Ware CF, Anderson G, Caamano JH (2010) Ontogeny of stromal organizer cells during lymph node development. J Immunol 184 (8):4521–4530. doi:jimmunol.0903113 [pii] 10.4049/jimmunol.0903113

Chen S-C, Leach MW, Chen Y, Cai X-Y, Sullivan L, Wiekowski M, Dovey-Hartman BJ, Zlotnik A, Lira SA (2002a) Central nervous system inflammation and neurological disease in transgenic mice expressing the CC chemokine CCL21 in oligodendrocytes. J Immunol 168 (3):1009–1017

Chen SC, Vassileva G, Kinsley D, Holzmann S, Manfra D, Wiekowski MT, Romani N, Lira SA (2002b) Ectopic expression of the murine chemokines CCL21a and CCL21b induces the formation of lymph node-like structures in pancreas, but not skin, of transgenic mice. J Immunol 168(3):1001–1008

Crowe PD, VanArsdale TL, Walter BN, Ware CF, Hession C, Ehrenfels B, Browning JL, Din WS, Goodwin RG, Smith CA (1994) A lymphotoxin-beta-specific receptor. Science 264 (5159):707–710

Cupedo T, Vondenhoff MF, Heeregrave EJ, De Weerd AE, Jansen W, Jackson DG, Kraal G, Mebius RE (2004) Presumptive lymph node organizers are differentially represented in developing mesenteric and peripheral nodes. J Immunol 173(5):2968–2975. doi:173/5/2968 [pii]

De Togni P, Goellner J, Ruddle NH, Streeter PR, Fick A, Mariathasan S, Smith SC, Carlson R, Shornick LP, Strauss-Schoenberger J et al (1994) Abnormal development of peripheral lymphoid organs in mice deficient in lymphotoxin. Science 264(5159):703–707

Dougall WC, Glaccum M, Charrier K, Rohrbach K, Brasel K, De Smedt T, Daro E, Smith J, Tometsko ME, Maliszewski CR, Armstrong A, Shen V, Bain S, Cosman D, Anderson D, Morrissey PJ, Peschon JJ, Schuh J (1999) RANK is essential for osteoclast and lymph node development. Genes Dev 13(18):2412–2424

Drayton DL, Liao S, Mounzer RH, Ruddle NH (2006) Lymphoid organ development: from ontogeny to neogenesis. Nat Immunol 7(4):344–353

Eberl G, Littman DR (2004) Thymic origin of intestinal alphabeta T cells revealed by fate mapping of RORgammat + cells. Science 305(5681):248–251. doi:10.1126/science.1096472 305/5681/248 [pii]

Eberl G, Marmon S, Sunshine MJ, Rennert PD, Choi Y, Littman DR (2004) An essential function for the nuclear receptor RORgamma(t) in the generation of fetal lymphoid tissue inducer cells. Nat Immunol 5(1):64–73. doi:10.1038/ni1022 ni1022 [pii]

Fan L, Reilly CR, Luo Y, Dorf ME, Lo D (2000) Cutting edge: ectopic expression of the chemokine TCA4/SLC is sufficient to trigger lymphoid neogenesis. J Immunol 164 (8):3955–3959

Forster R, Schubel A, Breitfeld D, Kremmer E, Renner-Muller I, Wolf E, Lipp M (1999) CCR7 coordinates the primary immune response by establishing functional microenvironments in secondary lymphoid organs. Cell 99(1):23–33

Furtado GC, Marinkovic T, Martin AP, Garin A, Hoch B, Hubner W, Chen BK, Genden E, Skobe M, Lira SA (2007) Lymphotoxin beta receptor signaling is required for inflammatory lymphangiogenesis in the thyroid. Proc Natl Acad Sci U S A 104(12):5026–5031. doi:10.1073/pnas.0606697104

Furtado GC, Pacer ME, Bongers G, Benezech C, He Z, Chen L, Berin MC, Kollias G, Caamano JH, Lira SA (2014) TNFalpha-dependent development of lymphoid tissue in the absence of RORgammat(+) lymphoid tissue inducer cells. Mucosal Immunol 7(3):602–614. doi:10.1038/mi.2013.79

Futterer A, Mink K, Luz A, Kosco-Vilbois MH, Pfeffer K (1998) The lymphotoxin beta receptor controls organogenesis and affinity maturation in peripheral lymphoid tissues. Immunity 9 (1):59–70. doi:S1074-7613(00)80588-9 [pii]

Goc J, Fridman WH, Sautes-Fridman C, Dieu-Nosjean MC (2013) Characteristics of tertiary lymphoid structures in primary cancers. Oncoimmunology 2(12):e26836. doi:10.4161/onci.26836

Kim D, Mebius RE, MacMicking JD, Jung S, Cupedo T, Castellanos Y, Rho J, Wong BR, Josien R, Kim N, Rennert PD, Choi Y (2000) Regulation of peripheral lymph node genesis by the tumor necrosis factor family member TRANCE. J Exp Med 192(10):1467–1478

Kojouharoff G, Hans W, Obermeier F, Mannel DN, Andus T, Scholmerich J, Gross V, Falk W (1997) Neutralization of tumour necrosis factor (TNF) but not of IL-1 reduces inflammation in chronic dextran sulphate sodium-induced colitis in mice. Clin Exp Immunol 107(2):353–358

Kong YY, Yoshida H, Sarosi I, Tan HL, Timms E, Capparelli C, Morony S, Oliveira-dos-Santos AJ, Van G, Itie A, Khoo W, Wakeham A, Dunstan CR, Lacey DL, Mak TW, Boyle WJ, Penninger JM (1999) OPGL is a key regulator of osteoclastogenesis, lymphocyte development and lymph-node organogenesis. Nature 397(6717):315–323. doi:10.1038/16852

Kontoyiannis D, Boulougouris G, Manoloukos M, Armaka M, Apostolaki M, Pizarro T, Kotlyarov A, Forster I, Flavell R, Gaestel M, Tsichlis P, Cominelli F, Kollias G (2002) Genetic dissection of the cellular pathways and signaling mechanisms in modeled tumor necrosis factor-induced Crohn's-like inflammatory bowel disease. J Exp Med 196(12):1563–1574

Lochner M, Ohnmacht C, Presley L, Bruhns P, Si-Tahar M, Sawa S, Eberl G (2011) Microbiota-induced tertiary lymphoid tissues aggravate inflammatory disease in the absence of RORgamma t and LTi cells. J Exp Med 208(1):125–134. doi:jem.20100052 [pii] 10.1084/jem.20100052

Lorenz RG, Chaplin DD, McDonald KG, McDonough JS, Newberry RD (2003) Isolated lymphoid follicle formation is inducible and dependent upon lymphotoxin-sufficient B lymphocytes, lymphotoxin beta receptor, and TNF receptor I function. J Immunol 170(11):5475–5482

Lovas A, Weidemann A, Albrecht D, Wiechert L, Weih D, Weih F (2012) p100 Deficiency is insufficient for full activation of the alternative NF-kappaB pathway: TNF cooperates with p52-RelB in target gene transcription. PLoS ONE 7(8):e42741. doi:10.1371/journal.pone.0042741 PONE-D-11-17269 [pii]

Luther SA, Lopez T, Bai W, Hanahan D, Cyster JG (2000) BLC expression in pancreatic islets causes B cell recruitment and lymphotoxin-dependent lymphoid neogenesis. Immunity 12(5):471–481. doi:S1074-7613(00)80199-5 [pii]

Luther SA, Bidgol A, Hargreaves DC, Schmidt A, Xu Y, Paniyadi J, Matloubian M, Cyster JG (2002) Differing activities of homeostatic chemokines CCL19, CCL21, and CXCL12 in lymphocyte and dendritic cell recruitment and lymphoid neogenesis. J Immunol 169(1):424–433

Marchesi F, Martin AP, Thirunarayanan N, Devany E, Mayer L, Grisotto MG, Furtado GC, Lira SA (2009) CXCL13 expression in the gut promotes accumulation of IL-22-producing lymphoid tissue-inducer cells, and formation of isolated lymphoid follicles. Mucosal Immunol 2(6):486–494

Marinkovic T, Garin A, Yokota Y, Fu Y-X, Ruddle NH, Furtado GC, Lira SA (2006) Interaction of mature CD3 + CD4+ T cells with dendritic cells triggers the development of tertiary lymphoid structures in the thyroid. J Clin Invest 116:2622–2632

Martin AP, Coronel EC, G-i S, Chen S-C, Vassileva G, Canasto-Chibuque C, Sedgwick JD, Frenette PS, Lipp M, Furtado GC, Lira SA (2004) A novel model for lymphocytic infiltration of the thyroid gland generated by transgenic expression of the CC chemokine CCL21. J Immunol 173:4791–4798

Martin AP, Canasto-Chibuque C, Shang L, Rollins BJ, Lira SA (2006) The chemokine decoy receptor M3 blocks CC chemokine ligand 2 and CXC chemokine ligand 13 function in vivo. J Immunol 177:7296–7302

Martin AP, Grisotto MG, Canasto-Chibuque C, Kunkel SL, Bromberg JS, Furtado GC, Lira SA (2008) Islet expression of M3 uncovers a key role for chemokines in the development and recruitment of diabetogenic cells in NOD mice. Diabetes 57(2):387–394

Mebius RE (2003) Organogenesis of lymphoid tissues. Nat Rev Immunol 3(4):292–303

Mebius RE, Rennert P, Weissman IL (1997) Developing lymph nodes collect CD4 + CD3- LTbeta + cells that can differentiate to APC, NK cells, and follicular cells but not T or B cells. Immunity 7(4):493–504

Miyawaki S, Nakamura Y, Suzuka H, Koba M, Yasumizu R, Ikehara S, Shibata Y (1994) A new mutation, aly, that induces a generalized lack of lymph nodes accompanied by immunodeficiency in mice. Eur J Immunol 24(2):429–434. doi:10.1002/eji.1830240224

Muniz LR, Pacer ME, Lira SA, Furtado GC (2011) A critical role for dendritic cells in the formation of lymphatic vessels within tertiary lymphoid structures. J Immunol. doi:10.4049/jimmunol.1004233

Neyt K, Perros F, GeurtsvanKessel CH, Hammad H, Lambrecht BN (2012) Tertiary lymphoid organs in infection and autoimmunity. Trends Immunol 33(6):297–305. doi:10.1016/j.it.2012.04.006

Okada T, Ngo VN, Ekland EH, Forster R, Lipp M, Littman DR, Cyster JG (2002) Chemokine requirements for B cell entry to lymph nodes and Peyer's patches. J Exp Med 196(1):65–75

Okuda M, Togawa A, Wada H, Nishikawa S (2007) Distinct activities of stromal cells involved in the organogenesis of lymph nodes and Peyer's patches. J Immunol 179(2):804–811. doi:179/2/804 [pii]

Picarella DE, Kratz A, Li CB, Ruddle NH, Flavell RA (1992) Insulitis in transgenic mice expressing tumor necrosis factor beta (lymphotoxin) in the pancreas. Proc Natl Acad Sci U S A 89(21):10036–10040

Randall TD, Carragher DM, Rangel-Moreno J (2008) Development of secondary lymphoid organs. Annu Rev Immunol 26:627–650. doi:10.1146/annurev.immunol.26.021607.090257

Rangel-Moreno J, Carragher DM, de la Luz G-HM, Hwang JY, Kusser K, Hartson L, Kolls JK, Khader SA, Randall TD (2011) The development of inducible bronchus-associated lymphoid tissue depends on IL-17. Nat Immunol 12(7):639–646. doi:10.1038/ni.2053 ni.2053 [pii]

Rennert PD, Browning JL, Mebius R, Mackay F, Hochman PS (1996) Surface lymphotoxin alpha/beta complex is required for the development of peripheral lymphoid organs. J Exp Med 184(5):1999–2006

Rennert PD, Browning JL, Hochman PS (1997) Selective disruption of lymphotoxin ligands reveals a novel set of mucosal lymph nodes and unique effects on lymph node cellular organization. Int Immunol 9(11):1627–1639

Rennert PD, James D, Mackay F, Browning JL, Hochman PS (1998) Lymph node genesis is induced by signaling through the lymphotoxin beta receptor. Immunity 9(1):71–79. doi:S1074-7613(00)80589-0 [pii]

Roozendaal R, Mebius RE (2011) Stromal cell-immune cell interactions. Annu Rev Immunol 29:23–43. doi:10.1146/annurev-immunol-031210-101357

Shang L, Thirunarayanan N, Viejo-Borbolla A, Martin A, Bogunovic M, Marchesi F, Unkeless J, Ho Y, Furtado G, Alcami A, Merad M, Mayer L, Lira S (2009) Expression of the chemokine binding protein M3 promotes marked changes in the accumulation of specific leukocytes subsets within the intestine. Gastroenterology 137:1006–1018

Sun Z, Unutmaz D, Zou YR, Sunshine MJ, Pierani A, Brenner-Morton S, Mebius RE, Littman DR (2000) Requirement for RORgamma in thymocyte survival and lymphoid organ development. Science 288(5475):2369–2373. doi:8641 [pii]

Tsuji M, Suzuki K, Kitamura H, Maruya M, Kinoshita K, Ivanov II, Itoh K, Littman DR, Fagarasan S (2008) Requirement for lymphoid tissue-inducer cells in isolated follicle formation and T cell-independent immunoglobulin A generation in the gut. Immunity 29(2):261–271. doi:S1074-7613(08)00303-8 [pii] 10.1016/j.immuni.2008.05.014

Viejo-Borbolla A, Martin AP, Muniz LR, Shang L, Marchesi F, Thirunarayanan N, Harpaz N, Garcia RA, Apostolaki M, Furtado GC, Mayer L, Kollias G, Alcami A, Lira SA (2010) Attenuation of TNF-driven murine ileitis by intestinal expression of the viral immunomodulator CrmD. Mucosal Immunol 3(6):633–644. doi:mi201040 [pii] 10.1038/mi.2010.40

Vondenhoff MF, Greuter M, Goverse G, Elewaut D, Dewint P, Ware CF, Hoorweg K, Kraal G, Mebius RE (2009) LTbetaR signaling induces cytokine expression and up-regulates lymphangiogenic factors in lymph node anlagen. J Immunol 182(9):5439–5445. doi:182/9/5439 [pii] 10.4049/jimmunol.0801165

Warnock RA, Campbell JJ, Dorf ME, Matsuzawa A, McEvoy LM, Butcher EC (2000) The role of chemokines in the microenvironmental control of T versus B cell arrest in Peyer's patch high endothelial venules. J Exp Med 191(1):77–88

Weih F, Caamano J (2003) Regulation of secondary lymphoid organ development by the nuclear factor-kappaB signal transduction pathway. Immunol Rev 195:91–105. doi:064 [pii]

Yokota Y, Mansouri A, Mori S, Sugawara S, Adachi S, Nishikawa S, Gruss P (1999) Development of peripheral lymphoid organs and natural killer cells depends on the helix-loop-helix inhibitor Id2. Nature 397(6721):702–706. doi:10.1038/17812

Chapter 7
Synthesis of Functional Tertiary Lymphoid Organs

Yuka Kobayashi, Koichi Kato, Makoto Nakamura, and Takeshi Watanabe

Abstract The regeneration of functional immune organ will be one of major focus in future immunology research. It will be an useful tool which induces efficient immune responses in the body on demand and offers effective ways to restore the immune status and treat uncontrollable obstinate diseases such as cancer, autoimmune diseases, severe infection and immuno-insufficiency/deficiency caused by tissue damages, abnormality, primary defect and aging. Artificially synthesized lymphoid organs may also provide us with a highly informative method not only for clinical aim but also basic study on the development and functions of immunological tissues and organs. We first reported successful generation of artificially-constructed lymph node-like tertiary lymphoid tissues at ectopic sites in mouse by applying certain stromal cell lines (Suematsu S, Watanabe T. Nat Biotechnol 22 (12):1539–1545, 2004; Okamoto N et al. J Clin Invest 117(4):997–1007, 2007; Kobayashi Y, Watanabe T. Trends Immunol 31(11):422–428, 2010). They showed a remarkable ability to induce immune responses upon antigen stimulation, especially when transplanted into naïve or immune-compromised hosts. In this review, we discuss about the rationale and method for the synthesis of functional tertiary lymph node-like lymphoid tissues in mouse. Especially, we discuss here on the method with applying only soluble factors but without using any stromal cell, that enables proper accumulation and functional organization of immune cells in grafts.

Y. Kobayashi • T. Watanabe (✉)
The Tazuke-Kofukai Medical Research Institute/Kitano Hospital, Kita-ku, Osaka 530-8480, Japan
e-mail: takeshi-watanabe@kitano-hp.or.jp

K. Kato
Department of Biomaterials Science, Graduate School of Biomedical Science, University of Hiroshima, Minami-ku, Hiroshima 734-8553, Japan

M. Nakamura
Faculty of Life Science and Engineering, University of Toyama, Gofuku 3190, Toyama 930-8555, Japan

T. Watanabe, Y. Takahama (eds.), *Synthetic Immunology*,
DOI 10.1007/978-4-431-56027-2_7

List of Abbreviations

SLO	secondary lymphoid organs
TLO	tertiary lymphoid organs
LN	lymph node
PP	Payer's patches
DC	dendritic cells
FDC	follicular dendritic cells
FRC	fibroblastic reticular cells
HEV	high endothelial venules
RIP- LTα	rat insulin promotor expressing *Lta* (lymphotoxin-α) gene
LTi	lymphoid tissue inducer
LTo	lymphoid tissue organizer
LTRβ	lymphotoxin receptor-beta
RA	retinoic acid
VEGFc	vessel endothelial growth factor
Aire	autoimmune regulator
MDSCs	myeloid-derived suppressor cells
Treg cells	regulatory T cells
HSPGs	heparan sulfate proteoglycans
HS	heparan sulfate
TES	tissue-engineered spleen
GC	germinal center
aAPC	artificial antigen-presenting cells
aLN	artificial lymph nodes
iPS	induced pluripotent stem cells

Introduction

The adaptive immune response induced in body upon encounter of antigens and pathogens plays a crucial role in the defense against infection and cancer. These adaptive immune responses also result in various outcomes such as immunological memory, immune tolerance, and sometime, pathological responses such as delayed hypersensitivity, allergy, inflammation or autoimmune diseases. These antigen-specific immune reactions are, in most cases, supported by secondary lymphoid organs (SLO) such as lymph nodes (LN), Peyer's patches (PP) and spleen, as well as by tertiary lymphoid organs (TLO). SLO develop during embryogenesis in the absence of antigen stimulation (Fu and Chaplin 1999; Mebius 2007; Roozendaal and Mebius 2011). TLO are formed in the postnatal period at sites of inflammation in any peripheral tissues/organs (Aloisi and Pujol-Borrell 2006). It is well known that the process of the TLO development partially recapitulates development of SLO. Both SLO and TLO structures include segregated T and B cell accumulations, $CD11c^{+}$ dendritic cell (DC) clusters, follicular dendritic cells (FDCs) networks and fibroblastic reticular cells (FRCs) networks, formation of high endothelial venules

(HEV) (Drayton et al. 2006; Thaunat et al. 2005; Krautler et al. 2012; Fletcher et al. 2011; Chai et al. 2013). TLO formation occurs at ectopic sites and often accompanies the onset of acute and chronic inflammation, infection, autoimmune diseases and tumors. TLO formation is also often associated with tissue damages (Drayton et al. 2006). It has been reported that TLO formation was observed in the various tissues such as synovial membrane of rheumatoid arthritis (Shi et al. 2001; Takemura et al. 2001; Young et al. 1984; Kobayashi et al. 2013), aorta adventitia of atherosclerosis (Grabner et al. 2009; Lotzer et al. 2010; Hu and Habenicht et al. 2015), salivary gland of Sjögren syndrome (Salomonsson et al. 2003; Stott et al. 1998), thymus in autoimmune Myasthenia Gravis (Thomas et al. 1982; Weiss et al. 2013), infection (Neyt et al. 2012), cancer (Dieu-Nosjean et al. 2014) and so on. Generation of TLO has also been reported during organ transplant rejection, which raises the possibility that TLO serve as a site for local adaptive immune responses and contribute to allograft rejection (Nasr et al. 2007), indicating that TLO formation may also serve as an effective immune organ/tissue. RIP-LTα transgenic mice, which express *Lta* (lymphotoxin-α) gene under control of the rat insulin promoter (RIP), developed TLO in the pancreatic islets, kidney and skin. When skin grafts of the RIP-LTα transgenic mouse (containing host TLO) were transplanted into allogeneic splenectomized aly/aly mice which lack all SLO, the skin graft with TLO containing host-derived lymphocytes was rejected (Nasr et al. 2007). Thus, the TLOs play a role as an activation site of naïve T cells to effecter and memory T cells. It has been recently reported that TLO in artery such as aorta could control aorta immunity and protect against atherosclerosis through lymphotoxin β receptor (LTβR) on vascular smooth muscle cells (Hu and Habenicht et al. 2015). Taken together, it is suggested that TLO might provide a highly localized immune response against microbial-derived antigens, tumor-derived antigens, auto- or allo-antigens, etc. These findings provide us with rationale for developing strategies to artificially synthesize tertiary lymphoid organs at ectopic local sites where strong and specific immune responses are desired. To synthesize functional artificial lymphoid tissues in vivo at ectopic sites or in vitro, the regulated processes of SLO and TLO formation must be precisely understood (Cupedo et al. 2012).

Roles of Lymphoid Tissue Inducer (LTi) Cells and Organizer (LTo) Cells in Development of SLO and TLO

The physiological processes that drive SLO development have been well studied (Fu and Chaplin 1999; Mebius 2007; Drayton et al. 2006; Ruddle and Akirav 2009). During embryogenesis, lymphoid tissue formation is initiated by the interaction between hematopoietic $CD45^{+}CD3^{-}CD4^{+}$ LT(lymphotoxin)α^{+} lymphoid-tissue inducer (LTi) cells and VCAM-1^{+}ICAM-1^{+} LTβR^{+} mesenchymal stromal lymphoid-tissue organizer (LTo) cells (Mebius 2007; Drayton et al. 2006; Randall et al. 2008; Lane et al. 2008; Mueller and Ahmed 2008; Benezech et al. 2010; Benezech et al. 2012). The LTi cells express LTα1β2 and also a variety of factors important for lymphoid organogenesis including the surface molecules IL-7Rα, RANK and transcription factors such as RORC and Id2 (Yokota et al. 1999; Kim et al. 2000; Kim et al. 2007; Eberl et al. 2004; Sun et al. 2000; Evans and Kim 2009; Kim et al. 2009). During the embryogenesis, LTi cells stimulate LTo cells through interaction of LTα1β2 on the LTi cell LTβR on the LTo surface with cell. LTβR ligation by LTα1β2 activates expression of the adhesion molecules such as VCAM-1, ICAM-1 and MAdCAM-1 on the LTo, as well assecretion of lymphoid chemokines CCL19, CCL21 and CXCL13 (Dejardin et al. 2002; Vondenhoff et al. 2009 Chemokines, CXCL13, CCL19, CCL21 and CXCL12, induce accumulation of immune cells such as B cells, T cells, and dendritic cells in lymphoid anlagen, which results in the formation of SLO. In human fetal tissues, LTi cells have been identified as lineage-negative RORC$^+$ IL-7Rα$^+$ hematopoietic cells. They are able to stimulate the expression of VCAM-1 and ICAM-1 on human mesenchymal LTo cells *in vitro*, through expression of LTα1β2 which triggers LTβR signaling in LTo cells (Cupedo et al. 2009).

LTi-like cells seem to be also involved in lymphoid organogenesis in the adult stage (Kim et al. 2007). It has been reported that adult LTi cells are recruited from bone marrow and maintained at sites of chronic inflammation or infection (Evans and Kim 2009). Adult LTi cells also appear to play a crucial role in the formation of TLO. Indeed, whole embryonic spleens implanted under the renal subcapsular space of adult mice induce lymphoid tissue formation through accumulation of host-derived T cells, B cells and DCs. Host LTi cells lead to stimulation of the grafted LTo cells and induce the production of CCL21 attracting naïve T cells and DCs (Glanville et al. 2009). The circulating adult LTi cells migrate into the transplanted embryonic spleen capsule grafts and interact with stromal cells to organize splenic white pulp formation (Glanville et al. 2009). These reports immediately suggested us that it might be possible to artificially generate lymphoid tissues in the adult body if we could provide proper LTo-like mesenchymal stromal cells or materials expressing the molecules, which substitute LTo function.

Mesenchymal Stromal (LTo) Cells in SLO and TLO

LTo cells are mesenchymal stromal cells and can be categorized according to their heterogeneous functions (Buettner et al. 2010). In SLO as well as TLO, it is well known that two important stromal cells play a specialized role in lymphoid tissue formation. They are the follicular dendritic cells (FDC) in B cell follicles (Krautler et al. 2012) and the fibroblastic reticular cells (FRC) in the T cell area (Fletcher et al. 2011; Chai et al. 2013). Their precise immunological contributions have been detailed in a review (Buettner et al. 2010). In the T cell area, FRC networks and their conduit system form three-dimensional structure that provide tracks for T cell migration, which is regulated by CCR7 and its ligands, CCL19 and CCL21. FRC are the source of survival factors for lymphocytes such as CCL19, IL-6 and IL-7, and B cell survival factor BAFF (Cremasco et al. 2014), as well as VEGF (Chyou et al. 2008) and retinoic acid (RA) (Hammerschmidt et al. 2008; Molenaar et al. 2009). RA leads to homing of activated T cells through activation of DCs (Iwata et al. 2004). VEGFc induces proliferation of endothelial cells and maintains blood vessels (Buettner et al. 2010). Furthermore, the production of VEGFc is induced in LTo cells, which contributes to lymphangiogenesis at lymph nodes (Vondenhoff et al. 2009). The conduit system is sheathed and interconnected by FRC (Katakai et al. 2004) and is also linked to lymphatic vessels with fluid flow at draining lymph nodes. It was shown that fluid flow in conduit, which is required for CCL21 secretion from FRC, enhanced the organization of FRC networks (Tomei et al. 2009), and that blocking lymph flow in peripheral lymph nodes decreased CCL21 and CCL19 gene expression in FRC. These data suggest that under conditions of increased lymph flow, for example, during inflammation, FRC networks increase in complexity, which could activate immune cell trafficking, lymph sampling and induction of immune responses. It has been reported that CD4 T cells play a central role in maintaining lymphoid tissue structure, especially FRC structure. CD4 depletion resulted in FRC loss because of lack of lymphotoxin-β produced by CD4 T cells (Zeng et al. 2012). LTo also play a crucial role in the formation of TLO. In inflammation foci, it has been shown that local resident fibroblasts (Peduto et al. 2009) and aortic smooth muscle cells (Grabner et al. 2009; Lotzer et al. 2010) give rise to LTo-like stromal cells, which express chemokines such as CXCL13, CCl19 and CCL21. Thus, the inflammation recapitulates SLO formation and results in the formation of TLO. Recently, CXCL13-secreting $CD4^+$ T cells have been identified in the human rheumatoid arthritis synovium. They may recruit CXCR5+ cells such as B cells and circulating follicular helper T cells into ectopic inflammation sites in synovium and play a role of formation of TLO (Kobayashi et al. 2013). These results suggest that not only mesenchymal stromal cells but also the immune cells such as $CD4^+$ T cells, B cells as well as dendritic cells may play a crucial role in the generation of TLOs.

Regulatory Role of LTo in Immune Responses

It was reported that lymph node FRC ectopically express and directly present peripheral tissue antigen (PTA) to naïve T cells under steady-state as well as inflammatory conditions (Fletcher et al. 2010). Another report also documented that lymph node-resident lymphatic endothelial cells directly present PTA to T cells and mediate peripheral tolerance independently of autoimmune regulator (Aire) (Cohen et al. 2010). It has been recently reported that many human cancers secrete CCL21 (Shields et al. 2010). Such CCL21-secreting tumors attract myeloid-derived suppressor cells (MDSCs) and regulatory T cells (Treg cells) besides naïve T cells into the tumor micro -environment, and also induce lymphoid-like stroma that equips the developing tumors with a substrate to promote the induction of Treg cells, which shift the host immune response from immunogenic to tolerogenic and facilitate tumor progression (Shields et al. 2010). These reports suggest that lymph node stromal cells (LTo) at adult stage appear to be involved in the maintenance of tolerance to self-antigens in the periphery, and that diverse function may be displayed by different stromal cell types. It has recently been reported that artery TLOs, which emerge in the aorta adventitia adjacent to atherosclerotic plaques, control aorta immunity and protect against atherosclerosis through vascular smooth muscle cell LTβR (Hu and Habenicht et al. 2015). Thus, artificially constructed TLO with different immunological functions including antigen-specific immune response, regulatory activity and protection of pathological tissue damages, may play an important role in the maintenance of homeostasis.

Contribution of Chemokines in Generation of TLO

As described above, both SLO and TLO possess organized structures, which include T and B cell areas, DCs, FDC networks and vascular networks. To initiate SLO and TLO development and to sustain their function, lymphoid chemokines play crucial roles in the migration and proper positioning of these lymphoid tissue-associated cells. In mice, CCL19 and CCL21 expressed by FRC and DCs attract and retain T cells within the appropriate area of the lymphoid tissues. Mice lacking both of CCL19 and CCL21, or mice lacking their receptor CCR7, retain structure of LNs and PPs but exhibit defects in T cell trafficking to these lymphoid tissues, and also in nasopharynx-associated lymphoid tissue maturation (Bai et al. 2000; Fukuyama et al. 2006). By performing immunohistochemical analysis of human SLO and TLO in inflammatory lesions, it was demonstrated that both of CCL19 and CCL21 are expressed by the smooth muscle actin-positive stromal cells, found in close proximity to HEVs. CCL21 is expressed also by mature DCs and lymphatic vessels (Manzo et al. 2007). CXCL12, the ligand for CXCR4, also contributes to T cell trafficking into LNs and PPs in collaboration with CCR7 ligands (Bai et al. 2000). Another important chemokine is CXCL13, which is also expressed by stromal cells including FDCs. CXCL13 initiates lymphoid organ formation (van de Pavert et al. 2009) and migration of CXCR5- expressing B cells into follicles, and increases expression of LTα by B cells. LTα further promotes FDC network development and CXCL13 expression, establishing a positive feedback loop of B cell follicle homeostasis (Ansel et al. 2000). Mice deficient for both CXCL13 and IL-7Rα lack most LNs, but CXCL13 is sufficient to recruit LTi cells, which express the CXCR5, to the ectopic sites in vivo Luther et al. 2003). These results, along with the IL-7Rα-mediated LTα1β2 upregulation on LTi cells, suggest that CXCL13 and IL-7Rα contribute to accumulation and activation of LTi cells, respectively, during the initial LTbR signaling-dependent stage in LN development. It has been shown, however, that in E12.5–14.5 LN anlagen, CXCL13 expression of LTo cells is induced by retinoic acid in a LTβR signaling-dependent manner (van de Pavert et al. 2009). In the SLO as well as TLO microenvironment, B cells circulate throughout B cell follicles to survey antigens. Once B cells engage antigens, the antigen-specific B cells increase CCR7 and CXCR5 expression, leading to their migration to the boundary between the B cell follicle and T cell zone, where B cells interact with antigen-specific T cells. Thus, controlled positioning of B cells in follicles is regulated by the finely-tune responsiveness of B cells to chemokines such as CXCL13, CCL19 and CCL21 (Reif et al. 2002). TLO formation was induced by ectopic expression of various chemokines, CXCL12, CXCL13, CCL19 and CCL21, as shown by chemokine-gene transgenic mouse (Fan et al. 2000; Hjelmstrom et al. 2000; Luther et al. 2002). Overexpression of each chemokine alone induced the complete formation of TLO. Ectopic expression of LTα/β or IL-7 also formed TLO Drayton et al. 2003; Meier et al. 2007). Thus, conditioning of these chemokines and cytokines into artificial scaffolds appears to be crucial for generation of artificial TLO.

Other Factors and Cells Required for Maintaining a Microenvironment for Lymphoid Tissue Organization

Ectopic lymphoid tissues known as inducible bronchus-associated lymphoid tissues (iBALT) were formed in influenza virus-infected mouse respiratory tracts through accumulation of lymphocytes and dendritic cells. Depletion of $CD11c^+$ DC from the lung leads to the gradual disappearance of $B220^+$ cells from the iBALT structure and a decrease of peanut agglutinin (PNA)-positive germinal center B cells, followed by diminution in the level of local immunoglobulin class switching. This suggests that $CD11c^+$ DCs support maintenance of iBALT structure and their function (GeurtsvanKessel et al. 2009), indicating that $CD11c^+$ DCs are also involved in TLO organogenesis in addition to LTo stromal cells. For proper lymphoid organogenesis during embryogenesis, spatiotemporal regulation of the cells and molecules that localize and function in the lymphoid anlagen appeared to be required. It has been reported that heparan sulfate proteoglycans (HSPGs) play a role in lymphoid development (Reijmers et al. 2010). HSPGs are molecules that contribute to several regulatory mechanisms during tissue morphogenesis by specifically binding soluble factors. In this regard, a number of lymphoid chemokines contain HSPGs binding motifs (Johnson et al. 2005; Rodgers et al. 2008). glucuronyl C5-epimerase mutant mice, which lack proper HS modifications, suffer from splenic hypoplasia, peripheral LN deletion, blood vessel mismatch-location and excess branching around peripheral lymph nodes (Reijmers et al. 2010). These results indicate that development of the peripheral lymphoid organs requires a tightly organized process that combines the signaling pathways for accumulation of hematopoietic cells, together with suitable mechanisms for retention of both lymphoid cells and bioactive molecules within the developing lymphoid organ. It will be important to take these findings into account for the synthesis of artificial TLO. It has been reported that CLEC-2, a natural ligand/

receptor for Gp38/Podoplanin, plays a role in lymphatic vasculature (Bertozzi et al. 2010). Recently, it was found that CLEC-2 is necessary for lymphatic cell ploriferation lymph node anlage persistence at birth. But, it also plays a crucial role in immune cell recirculation after repeated immunization and at chronic immune response (Benezech et al. 2014).

Scaffolds for Generation of Artificial Lymphoid Tissues

Tissue engineering has received much attention in the field of regenerative medicine and is undergoing continuous innovation. The characteristics of available biomaterials play an important role in tissue engineering strategies, and have allowed for the construction of a variety of tissues including blood vessels (Akita et al. 1997; Niklason et al. 1999), bone (Lutolf et al. 2003; Yu et al. 2004), spinal cord (Teng et al. 2002) and peripheral nervous system (Teng et al. 2002; Yu and Bellamkonda 2003). This, of course, is now being applied to lymphoid tissue formation. To maintain proper microenvironments at local sites, it will be necessary to apply scaffolds that are able to retain the appropriate immune cell populations, contribute to maintain the three dimensional structure of lymphoid organs, and produce or reserve soluble factors such as chemokines and cytokines expressed by stromal cells. A number of synthetic biomaterials have been developed which are able to duplicate the three dimensional microenvironments that are provided by natural extracellular matrices, such as fibrillar or non-fibrillar collagen, proteoglycans, matricellular proteins (Griffith and Swartz 2006; Lutolf and Hubbell 2005) and their hydrogels (Aberle et al. 2014). It has been demonstrated that structurally-engineered macroporous scaffolds, which combine poly ethylene glycol hydrogels with collagen, support T cell and DC migration (Stachowiak and Irvine 2008). Further advances are being made in this area (Griffith and Swartz 2006; Hori et al. 2009; Lutolf and Hubbell 2005). The first *in vivo* synthesis of artificial lymphoid tissues was achieved using a porous biocompatible collagen matrix, prepared from bovine Achilles tendon (referred to as a collagen sponge) (Suematsu and Watanabe 2004). This matrix has a non-homogeneous pore size ranging from 50 to 300 m. A difficulty in tissue engineering is achieving the ability to properly modulate or mimic dynamic tissue microenvironments, and to achieve for example appropriate cell recruitment or correct concentration gradients of soluble factors. As a first step toward this goal, biomaterials were recently described that are able to release several soluble factors, not only uniformly, but also gradually with temporal differences (Richardson et al. 2001). In this study, two growth factors were incorporated together into the same scaffolds by mixing PDGF-encapsulated polymer, lyophilized VEGF and non-treated polymer particles, so that the two growth factors were localized within distinct compartments. As a result, the scaffold could release the two factors with distinct kinetics and, after transplantation into subcutaneous tissue of Lewis rats, more mature vascular networks were structured than when using the scaffold which contained only a single growth factor (Richardson et al. 2001). Two other studies have also reported that sustained (Sun et al. 2005) or spatiotemporal (Chen et al. 2007) delivery of growth factors can lead to formation of a mature vascular system. Thus, in order to efficiently generate artificial organized tissues, the scaffold microenvironment should be carefully constructed to be as close as possible to the natural state.

Generation of Tissue-Engineered Spleen and Mucosal Lymphoid Tissues

Loss of spleen, which is caused by injury or surgical operation (splenectomy), is disadvantageous to certain bacterial infection. So far, tissue engineering for spleen regeneration has been most widely studied in a rodent model. Several reports have shown that not only embryonic and newborn spleen transplantation (Glanville et al. 2009; Willfuhr et al. 1992), but also tissue-engineered spleen (TES) generated from splenic multicellular components loaded on a biodegradable polymer scaffold (Grikscheit et al. 2008), induced partial or sufficient immune responses against infection. These studies provide valuable information concerning the generation of artificial lymphoid tissues. However, this approach lacks versatility in a clinical setting where the use of biological material might not be feasible if tissue supply is limited. Optimized combinations of stromal or hematopoietic cells and soluble factors such as chemokines or cytokines captured on scaffolds will be required to achieve solid results with general versatility. Recently, it has been reported a successful regeneration of mouse spleen possessing complete structure of both white pulp and red pulp by the purification of stromal cell population from newborn mouse and the injection of the cells into mouse renal subcapsular space (Tan and Watanabe 2014).

Artificial mucosal tissue with immune potential was developed by tissue-engineered neo-intestine synthesis (Perez et al. 2002). Neonatal 3-day-old rat bowel was first disaggregated and seeded onto biodegradable polymer tubes created from polyglycolic acid fibers coated with collagen. Constructs were then implanted into the omentum of adult recipients and, after 4 weeks, anastomosed to the native jejunum. After 20 weeks, the lymphoid tissue containing $CD3^{+}$T cells, $CD32^{+}$B cells, $CD56^{+}$NK cells, and $CD68^{+}$macrophages was detected in tissue-engineered neo-intestines comparable to that of native jejunum.

Generation of Artificial Lymph Node Tissues (aLN) in Mouse by Using Stromal Cells

To date, few studies have managed to combine structural formation of artificial lymphoid tissues with immune responsiveness. We hypothesized that since LTi may be supplied in the adult mouse, we might be able to generate transplantable lymphoid tissue in mouse if we could provide proper stromal cells and scaffolds. We applied, as a stromal cell, the thymus-derived stromal cell line TEL-2 (Nakashima et al. 1990; Kawabuchi et al. 1996), which was established in our laboratory. LTα cDNA was 12 transfected into the stromal cells. The LTα transfected stromal cells expressed VCAM-1 on the cell surface, and secreted various chemokines including CCL19, CCL21 and CXCL13. The mixtures of TEL-2 stromal cells and bone marrow-derived dendritic cells (BM-DCs) were injected into collagen sponges. Then, collagen sponges were transplanted into the mouse renal sub-capsular space. After 2–3 weeks, it was evident that grafts containing full of lymphocytes had formed on and in the collagen sponges. They possessed structures similar to secondary lymph node tissues, such as formation of segregated clusters of T and B cells, appearance of FDC and FRC networks in B cell and T cell areas, respectively, HEV-like structures at the endothelium of blood vessels, lymph vessel formation, and germinal center formation upon antigen stimulation. Thus, the grafts resembled secondary lymph nodes, and were termed as artificial lymph node tissue (Suematsu and Watanabe 2004). When they were formed in antigen-preimmunized mice, they could induce a strong secondary immune response *in vivo* upon antigen re-stimulation, and this high responsiveness was confirmed by the accumulation of memory and follicular helper T cells, as well as antigen-specific memory B cells (Okamoto et al. 2007). They could induce a strong immune response when transplanted into naïve mice and immunized with antigen. When they were re-transplanted into SCID mice and followed by immunization, a robust secondary immune response was induced which produced large amounts of antigen-specific high affinity IgG class antibody. This was the first success in generating artificial lymphoid tissue that was transplantable and immunologically functional. Recently, a novel in vitro culture system has been reported in which the germinal center (GC) reaction in mouse is reconstructed (Nojima et al. 2011). Mouse naïve B cells were cultured on the BALB/c 3T3 fibroblast cell line, which was stably transfected with CD40 ligand and BAFF together with cytokines, first IL-4 and then IL-21. In this culture system, GC-phenotype B cells underwent massive expansion and isotype switching. After in vivo adoptive transfer, they differentiated into memory type B cells and could elicit rapid secondary immune responses with the help of T cells.

Generation of Artificial Tertiary Lymphoid Tissues (TLOs) Without Using Stromal Cells (Figs. 7.1, 7.2, and 7.3)

According to the above information, it has been proven that application of stromal cell lines is a useful strategy for construction of artificial lymphoid tissues. However, these methods are subject to the limitation of generalized application for only a variety of recipient animals. In order to be versatile, we thought it is necessary to

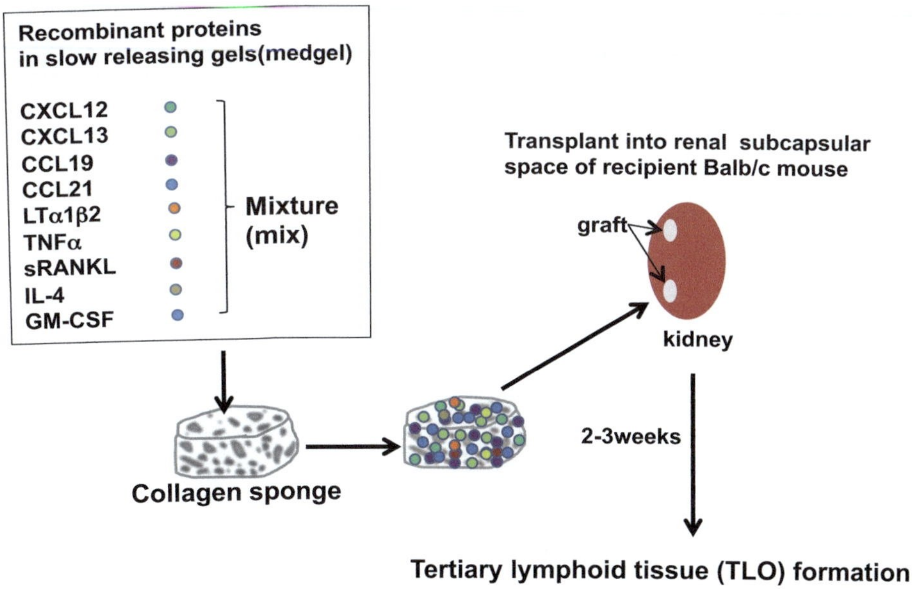

Fig. 7.1 Synthesis of functional TLO lymphoid tissues by cell-free system
The slow-releasing gels (medgels) containing each recombinant chemokine and cytokine protein were applied into the collagen sponge. Grafts were then transplanted into the renal subcapsular space of Balb/c mouse

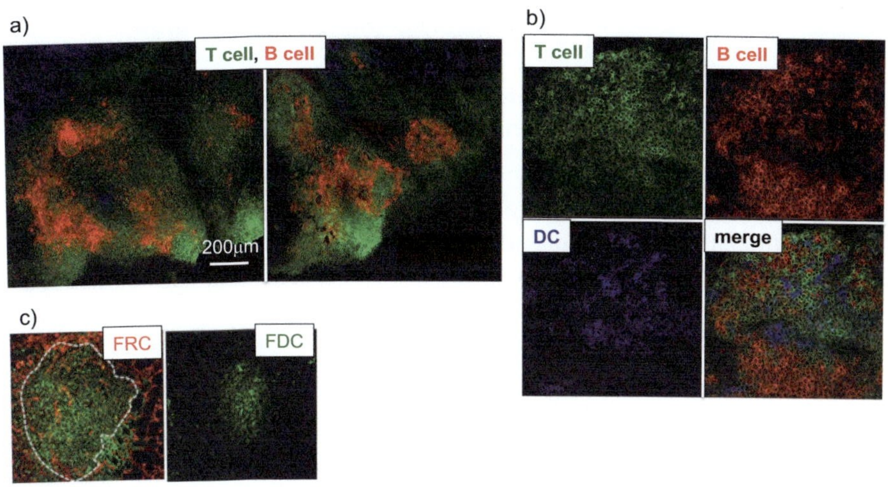

Fig. 7.2 Structures and immunological function of lymphoid tissues constructed by cell-free system
Three weeks after transplantation of the grafts, the artificial TLOs were formed. (**a**) T cell and B cell clusters and (**b**) DCs in T cell area in the TLO grafts. (**c**) FDC and FRC networks were also observed in the TLOs

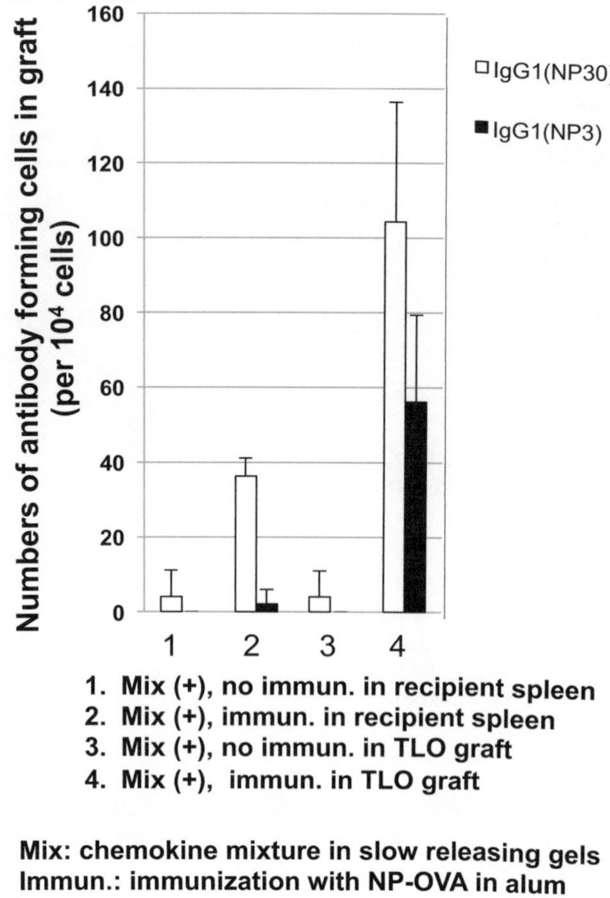

Fig. 7.3 Antigen-specific antibody formation in the TLO grafts
The NP-OVA primed-Balb/c mice harboring the artificial TLO grafts were intravenously immunized with the same antigen. Seven days after immunization, IgG1 class antigen-specific antibody-producing cells were detected in the TLO grafts. Higher numbers of IgG1 class antigen-specific, high affinity antibody-producing cells were detected within the grafts compared to the recipient spleen. *NP30* low affinity antibody, *NP3* high affinity antibody

introduce a new study for construction of artificial lymphoid tissues without applying stromal cells. Consequently, as an alternative, we are trying a cell-free method of transplanting collagen sponges containing lymphoid chemokines/cytokines and adhesion molecules (Fig. 7.1).

Firstly, recombinant chemokine and adhesion molecules were connected with a collagen-binding peptides (CBP) (Kalamajski et al. 2007) expecting that collagen sponge might be able to hold these molecules for a certain period. Recombinant chemokines, lymphotoxin-α1β2 and cytokines (TNFa, soluble RANKL (sRANKL), IL-4, GM-CSF) were included individually into the slow-releasing gels. The gels gradually release each protein for 4 weeks and resolved by collagenase. The gels were then injected into collagen sponges combined the chemokines and cytokines via CBP.

The collagen sponge grafts were then transplanted into the renal subcapsular space of mice. After 3 weeks, we have observed in these grafts the formation of large clusters of B cells and T cells, which were clearly segregated (Fig. 7.2a).

Many dendritic cells (DCs) were also detected in T cell area (Fig. 7.2b). In the graft, well-developed FDC and FRC networks were also observed (Fig. 7.2c). Mice carrying the grafts were immunized with antigen (NP-OVA). The significant numbers of IgG class high affinity antigen-specific antibody-producing cells were detected in the grafts (Fig. 7.3). While this method can be further improved, the artificially constructed TLOs by soluble factors could maintain the structure similar to SLO and function in vivo as an immune organ.

Attempts to Generate Human Peripheral Lymphoid Tissues

Generation of functional human-type artificial lymphoid tissues might be a noble and beneficial method for making advances in immunotherapy, as well as for investigating the physiological functions of human lymphoid tissues *in vivo*. For example, it is known in the tumor microenvironment that T cell immune responses are induced inside the tumor mass (Thompson et al. 2010). This, however, is not sufficient for tumor destruction due to immunosuppressive mechanisms. This immuno-suppression arises from an imbalance between activating and inhibitory signals that down-regulate immune cell function, as well as overproduction of immunosuppressive cytokines such as TGFβ, regulate immune cell function (Flavell et al. 2010). The atrophy of primary lymphoid organs, which occurs during aging and leads to immuno-senescence (Hadamitzky et al. 2010; Pan et al. 2008), gives rise to fatal infectious diseases. In these cases, generation of artificially-made lymphoid tissues might help supplement anti-tumor immune responses as well as be beneficial to boost immune activity and counteract immuno-suppression (Kobayashi and Watanabe 2010).

In terms of an *in vitro* attempt to create the human LN environment, it was demonstrated that antigen-specific B cell clusters could be observed in a bioreactor – an ex vivo culture device – containing a porous matrix with antigen-primed DCs (Giese et al. 2006). Long-term human peripheral lymphoid tissue cultures with immunologically competence in a 3D bioreactor have been reported (Kuzin et al. 2011). Potentially useful cell lines of artificial antigen-presenting cells (aAPC) were established for humans (Butler et al. 2007). These cells possess the capacity to stimulate *ex vivo* or *in vivo* production of cytotoxic T cells from naïve T cells (Butler et al. 2007). aAPC have also been made by covalently coupling antigen peptide-MHC complex and the B7.1 co-stimulatory molecule to magnetic beads. *In vivo* treatment with the aAPC induced significant or complete reduction of tumors in mice (Ugel et al. 2009). Based on this data, the aAPC might be an effective tool for tumor immunotherapy. The newly established aAPC might be also applicable for generating TLO. To apply artificial lymphoid tissues to human diseases, an *in vivo* human model would first have to be established, using knowledge from previous mouse experiments. There has been a much progress in the establishment and analysis of humanized mice (Ishikawa et al. 2005; Ito et al. 2002; Shultz et al. 2005; Traggiai et al. 2004; Shultz et al. 2010). NOD/SCID/IL-2 receptor γ chain null mice reconstituted with human $CD34^+$ hematopoietic stem cells develop a functional human hematopoietic and immune system. As a result, humanized mice are utilized as a human disease model. They can be experimentally manipulated and can be used to directly study infectious diseases, immunological disorders, as well as cancer therapy in humans in near future (Danner et al. 2011; Legrand et al. 2009; Saito et al. 2010; Shultz et al. 2007; Shultz et al. 2012; Zhang et al. 2007; Jaiswal et al. 2015). This model system could be a prime candidate for generating human artificial TLO. Recently, the successful generation of functional human organ buds (organoids) of human liver, kidney, pancreas, intestine, heart, lung and brain has been demonstrated from pluripotent stem cells (iPSc), by mixing with mesenchymal and endothelial cells (Takebe et al. 2013, 2015). These striking methods will be very useful also in regeneration of immunologically functional human TLO.

Conclusion

TLO are proved to be specialized lymphoid tissues that occasionally facilitate interaction between antigen presenting cells and effector lymphocytes for induction of proper adaptive and protective responses against pathological agents. The establishment of artificial TLO that function similarly to SLO, and that could represent a part of the regulatory immune reaction provides us with a novel strategy beneficial for both locally and systemic treatment of severe infection, autoimmune diseases, and cancer. From this point of view, development of the functional human type artificial tertiary lymphoid tissues or devices could be imperative in near future.

Acknowledgements This work was supported by the Grant-in-Aid for Scientific Research on Priority Areas from MEXT, Japan (Grant No. 24111009)

Disclosure

Authors report no conflicts of interest.

References

Aberle T, Franke K, Rist E et al (2014) Cell-type specific four-component hydrogel. PLoS One 9 (1):e86740

Akita M, Murata E, Merker HJ et al (1997) Morphology of capillary-like structures in a three-dimensional aorta/collagen gel culture. Ann Anat 179(2):127–136

Aloisi F, Pujol-Borrell R (2006) Lymphoid neogenesis in chronic inflammatorydiseases. Nat Rev Immunol 6(3):205–217

Ansel KM, Ngo VN, Hyman PL et al (2000) A chemokine-driven positive feedback loop organizes lymphoid follicles. Nature 406(6793):309–314

Bai Z, Hayasaka H, Kobayashi M et al (2000) CXC chemokine ligand 12 promotes CCR7-dependent naive T cell trafficking to lymph nodes and Peyer's patches. J Immunol 182(3): 1287–1295

Benezech C, White A, Mader E et al (2010) Ontogeny of stromal organizer cells during lymph node developmennt. J Immunol 184:4521–4530

Benezech C, Mader E, Desanti G et al (2012) Lymphotoxin-β receptor signaling through NF-κB2-RelB pathway reprogams adipocyte precursors as lymph node stromal cells. Immunity 37:1–14

Benezech C, Nayar S, Finney BA et al (2014) CLEC-2 is required for development and maintenance of lymph nodes. Blood 123(20):3200–3206

Bertozzi CC, Schmater AA, Mericko P et al (2010) Platelets regulate lymphatic vascular development through CLEC-2-SLP-76 signaling. Blood 116(4):661–670

Buettner M, Pabst R, Bode U (2010) Stromal cell heterogeneity in lymphoid organs. Trends Immunol 31(2):80–86

Butler MO, Lee JS, Ansen S et al (2007) Long-lived antitumor CD8+ lymphocytes for adoptive therapy generated using an artificial antigen-presenting cell. Clin Cancer Res 13(6):1857–1867

Chai Q, Onder L, Scandella E et al (2013) Maturation of lymph node fibroblastic reticular cells from myofibroblastic precuosors is critical for antiviral immunity. Immunity 38:1–12

Chen RR, Silva EA, Yuen WW et al (2007) Spatio-temporal VEGF and PDGF delivery patterns blood vessel formation and maturation. Pharm Res 24(2):258–264

Chyou S, Ekland EH, Carpenter AC (2008) Fibroblast-type reticular stromal cells regulate the lymph node vasculature. J Immunol 181(6):3887–3896

Cohen JN, Guidi CJ, Tewalt EF et al (2010) Lymph node-resident lymphatic endothelial cells mediate peripheral tolerance via Aire-independent direct antigen presentation. J Exp Med 207 (4):681–688

Cremasco V, Woodruff MC, Onder L et al (2014) B cell homeostasis and follicle confines are governed by fibroblastic reticular cells. Nat Immunol 15:973–981

Cupedo T, Crellin NK, Papazian N et al (2009) Human fetal lymphoid tissue-inducer cells are interleukin 17-producing precursors to $RORC^{+}$ $CD127^{+}$ natural killer-like cells. Nat Immunol 10(1):66–74

Cupedo T, Stroock A, Coles M et al (2012) Application of tissue engineering to the immune system: development of artificial lymph nodes. Front Immunol 3:343. doi:10.3389/fimmu.2012.00343

Danner R, Chaudhari SN, Rosenberger J et al (2011) Expression of HLA class II molecules in humanized NOD.Rag1KO.IL2RγcKO mice is critical for development and function of human T and B cells. PLoS One 6(5):e19826. doi:10.1371/journal.pone.0019826

Dejardin E, Droin NM, Delhase M et al (2002) The lymphotoxin-beta receptor induces different patterns of gene expression via two NF-kappaB pathways. Immunity 17(4):525–535

Dieu-Nosjean M-C, Goc J, Giraldo NA et al (2014) Tertiary lymphoid structures in cancer and beyond. Trends Immunol 35(11):571–580

Drayton DL, Ying X, Lee J (2003) Ectopic LT alpha beta directs lymphoid organ neogenesis with concomitant expression of peripheral node addressin and a HEV-restricted sulfotransferase. J Exp Med 197(9):1153–1163

Drayton DL, Liao S, Mounzer RH, Ruddle NH (2006) Lymphoid organ development: from ontogeny to neogenesis. Nat Immunol 7(4):344–353

Eberl G, Marmon S, Sunshine MJ et al (2004) An essential function for the nuclear receptor RORγ (t) in the generation of fetal lymphoid tissue inducer cells. Nat Immunol 5(1):64–73

Evans I, Kim MY (2009) Involvement of lymphoid inducer cells in the development of secondary and tertiary lymphoid structure. BMB Rep 42(4):189–193

Fan L, Reilly CR, Luo Y et al (2000) Cutting edge: ectopic expression of the chemokine TCA4/SLC is sufficient to trigger lymphoid neogenesis. J Immunol 164(8):3955–3959

Flavell RA, Sanjabi S, Wrzesinski SH et al (2010) The polarization of immune cells in the tumour environment by TGFbeta. Nat Rev Immunol 10(8):554–567

Fletcher AL, Lukacs-Kornek V, Reynoso ED et al (2010) Lymph node fibroblastic reticular cells directly present peripheral tissue antigen under steady-state and inflammatory conditions. J Exp Med 207(4):689–697

Fletcher AL, Malhotra D, Acton SE et al (2011) Reproducible isolation of lymph node stromal cells reveals site- dependent differences in fibroblastic reticular cells. Front Immunol. doi:10.3389/fimmu.2011.00035

Fu Y, Chaplin D (1999) Development and maturation of secondary lymphoid tissues. Annu Rev Immunol 17:399–433

Fukuyama S, Nagatake T, Kim DY et al (2006) Cutting edge: uniqueness of lymphoid chemokine requirement for the initiation and maturation of nasopharynx-associated lymphoid tissue organogenesis. J Immunol 177(7):4276–4280

Geurtsvankessel CH, Willart MA, Bergen IM et al (2009) Dendritic cells are crucial for maintenance of tertiary lymphoid structures in the lung of influenza virus-infected mice. J Exp Med 206(11):2339–2349

Giese C, Demmler CD, Ammer R et al (2006) A human lymph node in vitro – challenges and progress. Artif Organs 30(10):803–808

Glanville SH, Bekiaris V, Jenkinson EJ et al (2009) Transplantation of embryonic spleen tissue reveals a role for adult non-lymphoid cells in initiating lymphoid tissue organization. Eur J Immunol 39(1):280–289

Grabner R, Lotzer K, Dopping S et al (2009) Lymphotoxin beta receptor signaling promotes tertiary lymphoid organogenesis in the aorta adventitia of aged ApoE-/- mice. J Exp Med 206 (1):233–248

Griffith LG, Swartz MA (2006) Capturing complex 3D tissue physiology in vitro. Nat Rev Mol Cell Biol 7(3):211–224

Grikscheit TC, Sala FG, Ogilvie J et al (2008) Tissue-engineered spleen protects against overwhelming pneumococcal sepsis in a rodent model. J Surge Res 149(2):214–218

Hadamitzky C, Spohr H, Debertin AS et al (2010) Age-dependent histoarchitectural changes in human lymph nodes: an underestimated process with clinical relevance? J Anat 216 (5):556–562

Hammerschmidt SI, Ahrendt M, Bode U et al (2008) Stromal mesenteric lymph nodecells are essential for the generation of gut-homing T cells in vivo. J Exp Med 205(11):2483–2490

Hjelmstrom P, Fjell J, Nakagawa T et al (2000) Lymphoid tissue homing chemokines are expressed in chronic inflammation. Am J Pathol 156(4):1133–1138

Hori Y, Stern PJ, Hynes RO et al (2009) Engulfing tumors with synthetic extracellular matrices for cancer immunotherapy. Biomaterials 30(35):6757–6767

Hu D, Mohanta SK, Yin C et al (2015) Artery teriary lymphoid organs control aorta immunity and protect against atherosclerosis via vascular smooth muscle cell lymphotoxin β receptors. Immunity 42:1100–1115

Ishikawa F, Yasukawa M, Lyons B et al (2005) Development of functional human blood and immune systems in NOD/SCID/IL2 receptor {gamma} chain(null) mice. Blood 106(5): 1565–1573

Ito M, Hiramatsu H, Kobayashi K et al (2002) NOD/SCID/gamma(c)(null) mouse: an excellent recipient mouse model for engraftment of human cells. Blood 100(9):3175–3182

Iwata M, Hirakiyama A, Eshima Y et al (2004) Retinoic acid imprints gut-homing specificity on T cells. Immunity 21(4):527–538

Jaiswal S, Smith K, Ramirez A et al (2015) Dengue virus infection induces broadly cross-reactive human IgM antibodies that recognize intact virions in humanized BLT-NSG mice. Exp Biol Med 240(1):67–78

Johnson Z, Proudfoot AE, Handel TM (2005) Interaction of chemokines and glycosaminoglycans: a new twist in the regulation of chemokine function with opportunities for therapeutic intervention. Cytokine Growth Factor Rev 16(6):625–636

Kalamajski S, Aspberg A, Oldberg A (2007) The decorin sequence SYIRIADTNIT binds collagen type I. J Biol Chem 282(22):16062–16067

Katakai T, Hara T, Lee JH et al (2004) A novel reticular stromal structure in lymph node cortex; an immuno-platform for interactions among dendritic cells, T cells and B cells. Int Immunol 16 (8):1133–1142

Kawabuchi M, Nakamura K, Hirata K (1996) Morphological study of thymus stromal cells (TEL-2 cell) which play a role in the elimination of double positive immature thymocytes by phagocytosis. Anat Rec 244(3):271–283

Kim D, Mebius RE, Macmicking JD et al (2000) Regulation of peripheral lymph node genesis by the tumor necrosis factor family member TRANCE. J Exp Med 192(10):1467–1478

Kim MY, Mcconnell FM, Gaspal FM et al (2007) Function of CD4+CD3- cells in relation to B- and T-zone stroma in spleen. Blood 109(4):1602–1610

Kim MY, Kim KS, Mcconnell F, Lane P (2009) Lymphoid tissue inducer cells: architects of CD4 immune responses in mice and men. Clin Exp Immunol 157(1):20–26

Kobayashi Y, Watanabe T (2010) Synthesis of artificial lymphoid tissue with immunological function. Trends Immunol 31(11):422–428

Kobayashi S, Miura H, Shibuya H et al (2013) A distinct human CD4+ T cell subset that secretes CXCL13 in rheumatoid synovitis. Arthritis Rheum 65:3063–3072

Krautler NJ, Kana V, Kranich J et al (2012) Follicular Dendritic Cells emerge from ubiquitous perivascular precursors. Cell 150:194–206

Kuzin I, Sun H, Moshkani S et al (2011) Long-term immunologically competent human peripheral lymphoid tissue cultures in a 3D bioreactor. Biotechnol Bioeng 108:1430–1440

Lane P, Kim M-Y, Withers D et al (2008) Lymphoid tissue inducer cells in adaptive CD4 T cell dependent responses. Semin Immunol 20:159–163

Legrand N, Ploss A, Balling R et al (2009) Humanized mice for modeling human infectious disease: challenges, progress, and outlook. Cell Host Microbe 6(1):5–9

Lotzer K, Dopping S, Connert S et al (2010) Mouse aorta smooth muscle cells differentiate into lymphoid tissue organizer-like cells on combined tumor necrosis factor receptor-1/ lymphotoxin beta-receptor NF-kappaB signaling. Arterioscler Thromb Vasc Biol 30(3): 395–402

Luther SA, Bidgol A, Hargreaves DC et al (2002) Differing activities of homeostatic chemokines CCL19, CCL21, and CXCL12 in lymphocyte and dendritic cell recruitment and lymphoid neogenesis. J Immunol 169(1):424–433

Luther SA, Ansel KM, Cyster JG (2003) Overlapping roles of CXCL13, interleukin 7 receptor alpha, and CCR7 ligands in lymph node development. J Exp Med 197(9):1191–1198

Lutolf MP, Hubbell JA (2005) Synthetic biomaterials as instructive extracellular microenvironments for morphogenesis in tissue engineering. Nat Biotechnol 23(1):47–55

Lutolf MP, Weber FE, Schmoekel HG et al (2003) Repair of bone defects using synthetic mimetics of collagenous extracellular matrices. Nat Biotechnol 21(5):513–518

Manzo A, Bugatti S, Caporali R et al (2007) CCL21 expression pattern of human secondary lymphoid organ stroma is conserved in inflammatory lesions with lymphoid neogenesis. Am J Pathol 171(5):1549–1562

Mebius R (2007) Organogenesis of lymphoid tissues. Nat Rev Immunol 3(4):292–303

Meier D, Bornmann C, Chappaz S (2007) Ectopic lymphoid-organ development occurs through interleukin 7-mediated enhanced survival of lymphoid-tissue-inducer cells. Immunity 26(5): 643–654

Molenaar R, Greuter M, Van Der Marel AP et al (2009) Lymph node stromal cells support dendritic cell-induced gut-homing of T cells. J Immunol 183(10):6395–6402

Mueller SN, Ahmed R (2008) Lymphoid stroma in the initiation and control of immune responses. Immunol Rev 224:284–294

Nakashima M, Mori K, Maeda K, Kishi H et al (1990) Selective elimination of double-positive immature thymocytes by a thymic epithelial cell line. Eur J Immunol 20(1):47–53

Nasr IW, Reel M, Oberbarnscheidt MH et al (2007) Tertiary lymphoid tissues generate effector and memory T cells that lead to allograft rejection. Am J Transplant 7(5):1071–1079

Neyt K, Perros F, GeurtsvanKessel CH et al (2012) Tertiary lymphoid organs in infection and autoimmunity. Trends Immunol 33(6):297–305

Niklason LE, Gao J, Abbott WM et al (1999) Functional arteries grown in vitro. Science 284 (5413):489–493

Nojima T, Haniuda K, Moutai T et al (2011) In-vitro derived germinal centre B cells differentially generate memory B or plasma cells in vivo. Nat Commun. doi:10.1038/ncomms1475

Okamoto N, Chihara R, Shimizu C et al (2007) Artificial lymph nodes induce potent secondary immune responses in naive and immunodeficient mice. J Clin Invest 117(4):997–1007

Pan WR, Suami H, Taylor GI (2008) Senile changes in human lymph nodes. Lymphat Res Biol 6 (2):77–83

Peduto L, Dulauroy S, Lochner M et al (2009) Inflammation recapitulates the ontogeny of lymphoid stromal cells. J Immunol 182(9):5789–5799

Perez A, Grikscheit TC, Blumberg RS et al (2002) Tissue-engineered small intestine: ontogeny of the immune system. Transplantation 74(5):619–623

Randall TD, Carragher DM, Rangel-Moreno J (2008) Development of secondary lymphoid organs. Annu Rev Immunol 26:627–650

Reif K, Ekland EH, Ohl L et al (2002) Balanced responsiveness to chemoattractants from adjacent zones determines B-cell position. Nature 416(6876):94–99

Reijmers RM, Vondenhoff MF, Roozendaal R et al (2010) Impaired lymphoid organ development in mice lacking the heparan sulfate modifying enzyme glucuronyl C5-epimerase. J Immunol 184(7):3656–3664

Richardson TP, Peters MC, Ennett AB et al (2001) Polymeric system for dual growth factor delivery. Nat Biotechnol 19(11):1029–1034

Rodgers KD, San Antonio JD, Jacenko O (2008) Heparan sulfate proteoglycans: a GAGgle of skeletal-hematopoietic regulators. Dev Dyn 237(10):2622–2642

Roozendaal R, Mebius RE (2011) Stromal cell-immune cell interactions. Annu Rev Immunol 29:23–43

Ruddle NH, Akirav EM (2009) Secondary lymphoid organs: responding to genetic and environmental cues in ontogeny and the immune response. J Immunol 183(4):2205–2212

Saito Y, Uchida N, Tanaka S et al (2010) Induction of cell cycle entry eliminates human leukemia stem cells in a mouse model of AML. Nat Biotechnol 28(3):275–280

Salomonsson S, Jonsson MV, Skarstein K et al (2003) Cellular basis of ectopic germinal center formation and autoantibody production in the target organ of patients with Sjogren's syndrome. Arthritis Rheum 48(11):3187–3201

Shi K, Hayashida K, Kaneko M et al (2001) Lymphoid chemokine B cell-attracting chemokine-1 (CXCL13) is expressed in germinal center of ectopic lymphoid follicles within the synovium of chronic arthritis patients. J Immunol 166(1):650–655

Shields JD, Kourtis IC, Tomei AA et al (2010) Induction of lymphoidlike stroma and immune escape by tumors that express the chemokine CCL21. Science 328:749–752

Shultz LD, Lyons BL, Burzenski LM et al (2005) Human lymphoid and myeloid cell development in NOD/LtSz-scid IL2R gamma null mice engrafted with mobilized human hemopoietic stem cells. J Immunol 174(10):6477–6489

Shultz LD, Ishikawa F, Greiner DL (2007) Humanized mice in translational biomedical research. Nat Rev Immunol 7(2):118–130

Shultz LD, Saito Y, Najima Y et al (2010) Generation of functional human T-cell subsets with HLA-restricted immune responses in HLA class I expressing NOD/SCID/IL2r gamma (null) humanized mice. Proc Natl Acad Sci U S A 107(29):13022–13027

Shultz LD, Brehm MA, Garcia-Matinez JV et al (2012) Humanized mice for immune system investigation: progress, promise and challenges. Nat Rev Immunol 12:786–798

Stachowiak AN, Irvine DJ (2008) Inverse opal hydrogel-collagen composite scaffolds as a supportive microenvironment for immune cell migration. J Biomed Mater Res A 85(3): 815–828

Stott DI, Hiepe F, Hummel M et al (1998) Antigen-driven clonal proliferation of B cells within the target tissue of an autoimmune disease. The salivary glands of patients with Sjogren's syndrome. J Clin Invest 102(5):938–946

Suematsu S, Watanabe T (2004) Generation of a synthetic lymphoid tissue-like organoid in mice. Nat Biotechnol 22(12):1539–1545
Sun Z, Unutmaz D, Zou YR et al (2000) Requirement for RORgamma in thymocyte survival and lymphoid organ development. Science 288(5475):2369–2373
Sun Q, Chen RR, Shen Y (2005) Sustained vascular endothelial growth factor delivery enhances angiogenesis and perfusion in ischemic hind limb. Pharm Res 22(7):1110–1116
Takebe T, Sekine K, Enomura M et al (2013) Vascularized and functional human liver from an iPSC-derived organ bud transplant. Nature 499:481–485
Takebe T, Enomura M, Yoshizawa E et al (2015) Vascularized and complex organ buds from diverse tissues via mesenchymal cell-derived condensation. Cell Stem Cell 16:556–565
Takemura S, Braun A, Crowson C et al (2001) Lymphoid neogenesis in rheumatoid synovitis. J Immunol 167(2):1072–1080
Tan J, Watanabe T (2014) Murine spleen tissue regeneration from neonatal spleen capsule requires lymphotoxin priming of stromal cells. J Immunol 193:1194–1203
Teng YD, Lavik EB, Qu X et al (2002) Functional recovery following traumatic spinal cord injury mediated by a unique polymer scaffold seeded with neural stem cells. Proc Natl Acad Sci U S A 99(5):3024–3029
Thaunat O, Field AC, Dai J et al (2005) Lymphoid neogenesis in chronic rejection: evidence for a local humoral alloimmune response. Proc Natl Acad Sci U S A 102(41):14723–14728
Thomas JA, Willcox HN, Newsom-Davis J (1982) Immunological studies of the thymus in myasthenia gravis. Correction with clinical state and thymocyte culture responses. J Neuroimmunol 3(4):319–335
Thompson ED, Enriquez HL, Fu YX et al (2010) Tumor masses support naive T cell infiltration, activation, and differentiation into effectors. J Exp Med 207(8):1791–1804
Tomei AA, Siegert S, Britschgi MR et al (2009) Fluid flow regulates stromal cell organization and CCL21 expression in a tissue-engineered lymph node microenvironment. J Immunol 183(7): 4273–4283
Traggiai E, Chicha L, Mazzucchelli L et al (2004) Development of a human adaptive immune system in cord blood cell-transplanted mice. Science 304(5667):104–107
Ugel S, Zoso A, De Santo C et al (2009) In vivo administration of artificial antigen-presenting cells activates low-avidity T cells for treatment of cancer. Cancer Res 69(24):9376–9384
Van De Pavert SA, Olivier BJ, Goverse G et al (2009) Chemokine CXCL13 is essential for lymph node initiation and is induced by retinoic acid and neuronal stimulation. Nat Immunol 10(11): 1193–1199
Vondenhoff MF, Greuter M, Goverse G et al (2009) LTbetaR signaling induces cytokine expression and up-regulates lymphangiogenic factors in lymph node anlagen. J Immunol 182(9): 5439–5445
Weiss JM, Cufi P, Le Panse R et al (2013) The thymus in autoimmune Myasthenia Gravis: paradigm for a tertiary lymphoid organ. Rev Neurol (Paris) 169(8):640–649
Willfuhr KU, Westermann J, Pabst R (1992) Splenic autotransplantation provides protection against fatal sepsis in young but not in old rats. J Pediatr Surge 27(9):1207–1212
Yokota Y, Mansouri A, Mori S (1999) Development of peripheral lymphoid organs and natural killer cells depends on the helix-loop-helix inhibitor Id2. Nature 397:702–706
Young CL, Adamson TC 3rd, Vaughan JH et al (1984) Immunohistologic characterization of synovial membrane lymphocytes in rheumatoid arthritis. Arthritis Rheum 27(1):32–39
Yu X, Bellamkonda RV (2003) Tissue-engineered scaffolds are effective alternatives to autografts for bridging peripheral nerve gaps. Tissue Eng 9(3):421–430
Yu X, Botchwey EA, Levine EM et al (2004) Bioreactor-based bone tissue engineering: the influence of dynamic flow on osteoblast phenotypic expression and matrix mineralization. Proc Natl Acad Sci U S A 101(31):11203–11208
Zeng M, Palardini M, Engram JC et al (2012) Critical role of CD4 T cells in maintaining lymphoid tissue structure for immune cell homeostasis and reconstitution. Blood 120(9):1856–1867
Zhang L, Kovalev GI, Su L (2007) HIV-1 infection and pathogenesis in a novel humanized mouse model. Blood 109(7):2978–2981

Chapter 8
Growing a Surrogate Organ in Lymph Node

Eric Lagasse

Abstract Currently, the shortage of organs available for transplant into terminally ill patients represents a major worldwide medical, social, and economic challenge. For many years, the transplant of (stem) cells to regenerate failing organs has been proposed as an alternative to whole-organ transplantation. However, orthotopic cell-based therapy directed at a diseased organ may not be feasible for many patients. The efficacy of cell-based therapy becomes questionable in disease states which compromise the environment needed for cellular engraftment and function. This is true in many end stage organ failure scenarios such as cirrhotic or fibrotic liver. In another such example, DiGeorge syndrome patients suffer from an absence of thymic development causing severe immunodeficiency. In this instance, ectopic transplantation of allografted tissue is necessary to replace thymic function. Consequently, a critical component of cell therapy for these patients is the establishment of an optimal *in vivo* ectopic cell or tissue transplant site to achieve restoration of organ function. We have pioneered use of the lymph node as a site for ectopic tissue and organ development. In this chapter, we will discuss two applications, the transplantation of hepatocytes and thymic tissues in lymph node, to develop functional ectopic organs.

List of Abbreviations

OLT	Orthotopic liver transplantation
Fah	Fumarylacetoacetate hydrolase
NTBC	2-(2-nitro-4-trifluoro-methylbenzyol)-1,3-cyclohexanedione
IP	intraperitoneally
SP	splenic injection
BrdU	5-Bromo-2′-deoxyuridine
CK8	cytokeratin 8
CK5	cytokeratin 5
ES/iPSC	Embryonic Stem Cells/Induced Pluripotent Stem Cells

E. Lagasse (✉)
McGowan Institute for Regenerative Medicine, Department of Pathology, University of Pittsburgh School of Medicine, Pittsburgh, PA 15219, USA
e-mail: Lagasse@Pitt.edu

T. Watanabe, Y. Takahama (eds.), *Synthetic Immunology*,
DOI 10.1007/978-4-431-56027-2_8

Growing a Surrogate Liver in a Lymph Node

Problems Facing Patients with Liver Disease

In recent years, chronic liver diseases and cirrhosis are responsible for over 30,000 deaths annually in the United States alone. Orthotopic liver transplantation (OLT) is too often the last resort and is currently the only curative treatment for patients suffering from these severe hepatic diseases (Starzl et al. 1968; Reuben 2007). In addition, liver transplantation from one individual to another has severe limitations. Just over 6000 patients are transplanted each year (USA alone) out of an approximate 20,000 on the waiting list with an estimated 100,000 patients in need of a new liver. The shortage of available donors and unwanted immunologic rejections are two of the major challenges facing patients affected by end-stage liver diseases. Furthermore, patients with co-morbidities and advanced age are either not considered candidates for OLT or are expected to have reduced post-transplant survival (Lipshutz and Busuttil 2007; Volk et al. 2007; Perkins et al. 2009). A final additional concern is the cost of these dramatic procedures. For these reasons, cell-based transplantation has been proposed as either a therapeutic alternative to OLT or as a bridge for patients waiting for an available organ (Strom et al. 1999). To date, hepatocyte transplantation has demonstrated its functional utility in animal models. From the transgenic urokinase (Rhim et al. 1994; Braun and Sandgren 2000; Weglarz et al. 2000) to the induced tyrosinemic models (Grompe et al. 1995; 1998; Overturf et al. 1996, 1997; Lagasse et al. 2000; Grompe 2006) in mice, hepatocyte transplantation has successfully established its therapeutic potential achieving nearly complete liver regeneration by the transplanted cells. In spite of this encouraging prospect, human hepatocyte transplantation is still an experimental procedure undergoing clinical exploration (Fox et al. 1998; Strom et al. 1999) in the hope that translation of animal studies, and their therapeutic utility to human disorders, is feasible.

For patients suffering from end-stage liver disease, there is an additional challenge: most future cellular therapies have been directed at cell engraftment in the terminally diseased liver. Transplanted liver cells are generally injected *via* the spleen (splenic artery in patients or splenic parenchyma in rodents) or *via* the portal vein. Liver cells rapidly migrate, actively or passively, to the diseased liver, where hepatic regeneration is expected to occur. This approach limits, and possibly prevents, the effectiveness of cellular therapy in a vast majority of these patients due to the presence of cirrhotic and fibrotic liver tissue, common pathological features of late-stage liver disease. Therefore, liver regeneration at this point is a major challenge. In summary, patients suffering from terminal liver diseases have very few options available either currently or in the foreseeable future. The development of new therapeutic applications for this patient population is crucial to meet the continuing substantial demand and to provide potential therapeutic benefits.

Intraperitoneal Injection of Hepatocytes Rescues a Mouse Model of Lethal Liver Disease

To explore the feasibility and potential location of a functional ectopic liver, we used a mouse model of hereditary tyrosinaemia type I ($Fah^{-/-}$), a model of high-efficiency liver regeneration (Grompe et al. 1995; Overturf et al. 1997). $Fah^{-/-}$ tyrosinemic mice have progressive and fatal liver failure unless treated with 2-(2-nitro-4-trifluoro-methylbenzyol)-1,3-cyclohexanedione (NTBC, nitisone, Orphadin®) (Grompe et al. 1995). Our lab and others have shown that wild type hepatocytes have a strong selective growth advantage when transplanted in the livers of $Fah^{-/-}$ mice following NTBC removal, with near-complete regeneration of the native liver (Overturf et al. 1997; Lagasse et al. 2000). To evaluate ectopic locations for transplant, liver cells from congenic wild type mice were transplanted in $Fah^{-/-}$ mice subcutaneously, under the kidney capsule or intraperitoneally (IP), with splenic injection (SP) used as a positive control. The mice were taken off NTBC, inducing progressive liver failure in all animals, and weighed weekly to monitor liver function. Mice treated with IP injection initially lost weight and then spontaneously regained weight with donor hepatocytes repopulating the diseased liver and reversing lethal tyrosinemia (Hoppo et al. 2011). Subcutaneous injections and injections under the kidney capsule failed to rescue $Fah^{-/-}$ mice under similar conditions. However, IP transplantation of liver cells resulted in long-term survival of $Fah^{-/-}$ mice, suggesting that liver functions were restored. We found that donor hepatocytes rapidly colonized the local lymphatic system instead of the liver and rescued the animals from fatal liver disease (Hoppo et al. 2011). Within two days of IP transplantation, hepatocytes were detected adjacent to the lymphoid cells in the subcapsular sinus of some lymph nodes located along the gastric and common hepatic arteries. Two weeks after transplantation, hepatocytes had entirely colonized the lymph nodes (Hoppo et al. 2011). This outcome suggests that hepatocytes rapidly migrate into the lymphatic system through afferent lymph vessels, proliferate, and then passively or actively eliminate lymphocytes from the lymph nodes. Ten weeks post transplantation, after an apparent reversal of tyrosinemia, laparotomies were performed on the mice. Numerous enlarged nodules were observed around the stomach region and along the mesentery in all the animals transplanted by IP injection (Hoppo et al. 2011). None of these enlarged nodules were found in SP transplanted $Fah^{-/-}$ mice. Immunohistological analyses confirmed the presence of healthy donor hepatocytes in all the analyzed enlarged nodules. The newly generated "extra-hepatic nodules" had a mass of ectopic liver cells representing over 70 % of expected normal liver mass with the number of hepatocytes estimated at 1.5×10^7 (Hoppo et al. 2011). Analysis of the hepatized lymph nodes showed that not only had the lymphocytes almost completely disappeared, but high endothelial venules, the specialized post-capillary venules found in lymphoid tissue, were also absent after hepatocyte colonization (Hoppo et al. 2011). The high endothelial venules were replaced with large vessels that have a histology and size similar to that found in normal liver. These vessels appeared to be abundant in hepatized lymph nodes, indicating a possible adaptation of the vasculature to the newly generated hepatic tissue. The presence of the "extra-hepatic nodules" and the reversal of tyrosinemia was a long-lasting effect. Nine months after transplantation, animals were still alive, with no sign of liver disease and metabolic disorder, due to the presence of the ectopic livers (Hoppo et al. 2011).

Targeting the Lymph Node for Transplantation

In order to translate the functional reconstitution of $Fah^{-/-}$ mice to a clinical setting, several issues must be resolved. Hundreds of lymph nodes are distributed throughout the human or mouse body, collectively forming the lymphatic system. Restricting the development of an ectopic liver to a few specific lymph nodes is necessary to localize and possibly control the growth of ectopic tissues. To test this approach, a single mesenteric lymph node was injected in $Fah^{-/-}$ mice with 10^5 wild type hepatocytes. The jejunal lymph node was chosen for the site of injection because it is the largest lymph node in the mouse. Using *in vivo* imaging, syngeneic luciferase-expressing hepatocytes transplanted directly into the jejunal lymph node showed that hepatocytes remain at this site for at least a week. In contrast, splenic and intraperitoneal injection of hepatocytes resulted in the rapid dispersion of cells to the liver and throughout the abdominal cavity, respectively. Ten days after removal of NTBC, a cluster of Fah^{+} hepatocytes was identified in the injected node. Three months later and after two cycles of selection (NTBC removal), recipient mice were rescued from tyrosinemia. A single large hepatic mass (ectopic liver) was observed at the injection site, originally the jejunal lymph node. Histology and Fah staining confirmed the presence of donor hepatocytes. In subsequent studies, we also showed that injection of a single peripheral lymph node (a popliteal, peri-portal, or an axillary lymph node) resulted in similar outcomes, ectopic growth of liver tissue and rescue of the animals from lethal liver disease (Komori et al. 2012).

In all the experiments so far, syngeneic donor hepatocytes were used to transplant cells into recipient lymph nodes. Because lymph nodes have an essential role in the immune system for mounting an allogeneic response (Lakkis et al. 2000), we also tested whether successful engraftment of allogeneic hepatocytes in lymph nodes is possible. Allogeneic hepatocytes were transplanted into a single lymph node and the results were compared to animals treated with splenic injection. Whereas allogeneic hepatocytes transplanted into untreated $Fah^{-/-}$ recipients mice did not engraft, blocking the co-stimulation pathways CD28-B7 and CD40-CD40L in recipient mice facilitated successful engraftment of allogeneic hepatocytes in both control and experimental animals. This experiment suggested there is no specific barrier in lymph nodes preventing the engraftment of allogeneic cells. Therefore, direct transplantation of hepatocytes into a single lymph node is a promising approach to site-specific ectopic liver generation.

Conclusion

Our data demonstrates transplantation of hepatocytes in lymph node will generate a life-sustaining functional ectopic liver in a clinically relevant animal model of hepatic disease. Lymph nodes might be amenable to hepatic engraftment because of their high degree of vascularization. It has been reported that an inadequate vascular supply leads to hepatocyte death due to hypoxia within a few days of ectopic transplantation (Henne-Bruns et al. 1991; Gupta et al. 1994; Smith and Mooney 2007). Since lymph nodes are well vascularized for support of lymphocyte activity (von Andrian 1996), they may be better suited to the immediate survival of engrafting hepatocytes. Furthermore, our data supports the model that proliferation of donor hepatocytes in the lymph nodes is homeostatically controlled. BrdU labeling experiments indicate that donor hepatocytes proliferate for 2–3 weeks post-injection in the lymph nodes and cease to proliferate by 8 weeks after injection (data not published). Therefore, once the hepatocytes in ectopic sites attain a critical mass, they cease to proliferate.

Based on the promising results with hepatocytes in lymph nodes, we have proposed transplanting other cells and tissues to test for the functional development of their respective organs. Broadening the concept to grow surrogate organs in lymph nodes could be applied to a wide-ranging area of biomedical needs. And probably one of the most interesting organs to generate in a lymph node is the thymus.

Growing a Surrogate Thymus in a Lymph Node

Therapeutic Advantages of Growing a Surrogate Thymus

Transplanting thymic functions is of great therapeutic interest for the simple reason that the thymus is a central organ of our immune system. The thymus plays a crucial role by supporting the development of functional T cells. It is also the main organ involved in establishing immune tolerance through the elimination of auto-reactive T cell subsets. Both of these critical functions are essential for survival, with healthy individuals balancing self and non-self discrimination between the extremes of immunodeficiency and autoimmunity. Furthermore, because thymic functions can be compromised and lost due to aging, stress, irradiation, or chemotherapy, restoring a functional thymus after these conditions would be advantageous.

Remodeling a Thymus in a Lymph Node

A critical component in restoring thymic function is the establishment of an optimal *in vivo* site for thymus transplantation. Thymus transplantation faces a host of challenges, such as achieving adequate vascularization and function of the implanted cells, as well as avoiding immune rejection. We have shown that many of these challenges can be overcome by direct injection of donor cells, like hepatocytes, into lymph nodes. Our study reveals a remarkable capacity of lymph nodes to support the growth of complex organs, indicating transplantation of ectopic thymus into lymph node is feasible and representing an interesting approach to regenerative medicine (Weir 2012; Zakrzewski et al. 2012). Furthermore, lymph nodes could provide other, currently unidentified, advantages compared to other proposed ectopic transplant sites for the thymus due to their role as secondary lymphoid organs.

In the past, efforts to transplant thymic tissue into highly vascularized anatomical sites have shown clinical success in patients with DiGeorge syndrome, a genetic condition characterized by an absent or hypoplastic thymus (Markert et al. 2010; Davies 2013). In addition, artificial microenvironments for T-cell production have been fabricated for *in vitro* and *in vivo* studies (Poznansky et al. 2000; Rodewald 2000; Gill et al. 2002; Clark et al. 2005; Seach et al. 2010). Yet all of these strategies to restore thymic functions still have shortcomings including the limited viability of transplanted cells, an inadequate supply of donor cells, and diverse safety concerns. On the other hand, lymph nodes appear as ideal sites for ectopic thymic growth because they are highly vascularized lymphoid niches. To support this point of view, lymph nodes have long been known as 'secondary' lymphoid organs where mature lymphocytes interact with antigen-presenting cells. Upon their encounter with cognate antigen, the mature lymphocytes undergo activation and proliferation. Maturation of lymphocytes normally occurs in 'primary' lymphoid organs. B cells develop in the bone marrow and T cells in the thymus. Yet previous observations have pointed to the possibility that, in some scenarios, lymph nodes can support primary lymphocyte development. For example, during periods of lymphopenia shortly after bone marrow transplantation, the secondary lymphoid organs, including lymph nodes, support T cell development (Terra et al. 2005; Maillard et al. 2006). That said *de novo* generation of functional T cells from donor progenitor cells presents a unique challenge.

Rebuilding a Thymus in a Nude Mouse

To demonstrate that a functional ectopic thymus in a lymph node is feasible, we utilized the athymic nude mouse, a mouse model with a spontaneous $Foxn1^{nu}$ genetic mutation that causes defective development of the thymic epithelium, resulting in the absence of a thymus, and ultimately T-cell immunodeficiency. Using an approach similar to that used in the production of an ectopic functional liver in a tyrosinemic mouse, we asked whether a functional thymus could be generated in a lymph node of an athymic nude mouse. C57bl/6 GFP+ newborn thymuses were minced and injected in a mesenteric lymph node of an adult nude mouse using a 26-gage needle. Our team showed that direct injection of thymic tissue from a donor control mouse into a single lymph node resulted in ectopic growth and survival of a functional thymus (Komori et al. 2012). The thymic transplant resulted in circulating recipient single-positive CD4+ and CD8+ T cells (Komori et al. 2012). Ten months after transplantation, single-positive CD4+ and CD8+ T cells were still present in the peripheral blood, indicating long-term thymic engraftment in the lymph node (Komori et al. 2012). We then harvested the ectopic thymuses for further characterization of the engraftment. GFP+ epithelial thymic cells remained within the injected lymph node and were organized into an epithelial thymic structure (Komori et al. 2012). Previously, thymic epithelia have been distinguished by their cytokeratin 5 (CK5) and cytokeratin 8 (CK8) phenotypes in the thymic medulla and cortex, respectively. The ectopic thymuses present in the lymph node subcapsular sinus contained a mosaic of medulla (CK5+) and cortex (CK8+) regions (Komori et al. 2012). We also analyzed the ectopic thymuses for the presence of recipient double-positive CD4+CD8+ thymocytes, representing immature T cells undergoing thymic selection. The ectopic thymus contained recipient double-positive thymocytes as well as single-positive CD4+ and CD8+ T cells, indicating a functional selective mechanism of T-cell commitment and maturation (Komori et al. 2012). Importantly, metabolic activity, homeostatic properties and cellular functions of the transformed lymph node resembled those of the donor thymic tissue. For example, lymph node injected with thymic tissue gave rise to a normal T-cell receptor repertoire; i.e. regulatory, central memory, and effector memory T-cells; as well as new T-cells capable of rejecting skin graft or xenogeneic tumors (Komori et al. 2012). As expected, donor cells injected into lymph node of allogeneic mice were rejected unless the recipient mice were concomitantly treated with immunosuppressive agents (Komori et al. 2012).

T cells undergo a highly regulated 'education' processes during thymic development. The resulting thymic output consists of naive T cells that are able to confer immunity to 'non-self' while avoiding autoimmunity, characterized by the unwanted attack on healthy 'self' tissues by T cells. To determine whether T cells produced in lymph nodes injected with allogeneic thymic tissue are tolerant to self or donor tissue, Balb/c nude recipient mice ($H2^d$) were transplanted with fully allogeneic ($H2^b$) C57bl/6 thymuses in lymph node. We then carried out skin grafting experiments and showed that C57bl/6 skin grafts were accepted in the Balb/c nude recipients, while ($H2^k$) CBA/CaJ skin grafts were rejected. These results suggest that allogeneic donor thymic tissue, enriched in epithelial cells, when transplanted into a lymph node, supports the development of functional T-cells of recipient origin that are tolerant to both recipient and donor tissue. Although our published study does not fully elucidate the mechanism of tolerance induction to donor tissue (Komori et al. 2012), these important observations nevertheless indicate that this method could have potential as a strategy for inducing tolerance to alloantigens, a major point of interest in the field of solid-organ transplantation.

Synopsis

From a practical point of view, lymph nodes are an attractive target for cell therapy because they are easy to locate and access. Injecting a patient's lymph node with cells would likely be a minimally invasive and relatively safe procedure. The human body contains about 600 lymph nodes. Removal of a single lymph node or a small number of lymph nodes will cause minimal damage to lymphatic vessels and would not be expected to result in significant lymphedema or other local complications. Moreover, we show previously that the potential loss of the immune function of a small number of lymph nodes following cell or tissue transplantation did not compromise overall immunity (Komori et al. 2012). Importantly, our study found that overcoming immune rejection of transplanted cells was not more challenging when using lymph node as compared with other transplantation sites (Komori et al. 2012).

Interestingly, with regard to the need for immunosuppressive therapy in the scenario of allogeneic transplantation barriers, recent advances in embryonic stem cells and induced pluripotent stem cells (ES/iPSC) technology are making transplantation of custom-made, autologous iPSC-derived thymic epithelial cells and tissues possible (Parent et al. 2013), thereby overcoming or at least lowering immunological barriers to transplantation and again raising the question about the best possible transplantation site to generate a functional thymus with the ES/iPSC technology.

Overall, our work suggests that lymph node can be manipulated by transplantation of cells of interest and that lymph node function can be redirected from facilitating lymphocyte activation to ectopic organogenesis. We have already shown that even more tissues than liver and thymus can grow in lymph nodes (DeWard et al. 2014; Francipane and Lagasse 2014, 2015a, b). The remarkable plasticity and excellent vascularization of the lymph nodes, combined with the simplicity of the transplantation procedure, have significant implications for clinical translation and will likely encourage the development of new treatment protocols (Fig. 8.1).

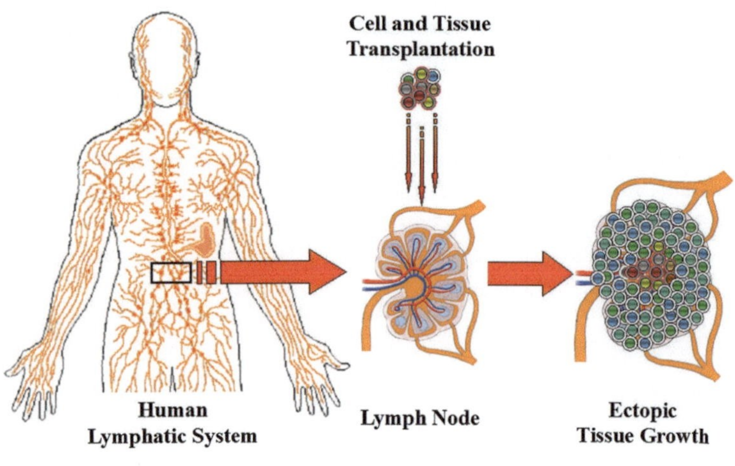

Fig. 8.1 Proposed approach to rebuild an organ in patient using lymph node as a site for regeneration

Acknowledgments This work was supported by the NIH Grant R01-DK085711. We thank Lynda Guzik for proofreading and editing.

References

Braun KM, Sandgren EP (2000) Cellular origin of regenerating parenchyma in a mouse model of severe hepatic injury. Am J Pathol 157(2):561–569

Clark RA, Yamanaka K, Bai M, Dowgiert R, Kupper TS (2005) Human skin cells support thymus-independent T cell development. J Clin Invest 115(11):3239–3249

Davies EG (2013) Immunodeficiency in DiGeorge syndrome and options for treating cases with complete athymia. Front Immunol 4:322

DeWard AD, Komori J, Lagasse E (2014) Ectopic transplantation sites for cell-based therapy. Curr Opin Organ Transplant 19(2):169–174

Fox IJ, Chowdhury JR, Kaufman SS, Goertzen TC, Chowdhury NR, Warkentin PI, Dorko K, Sauter BV, Strom SC (1998) Treatment of the Crigler-Najjar syndrome type I with hepatocyte transplantation. N Engl J Med 338(20):1422–1426

Francipane MG, Lagasse E (2014) Maturation of embryonic tissues in a lymph node: a new approach for bioengineering complex organs. Organogenesis 10:323–331

Francipane MG, Lagasse E (2015a) The lymph node as a new site for kidney organogenesis. Stem Cells Transl Med 4(3):295–307

Francipane MG, Lagasse E (2015b) Pluripotent stem cells to rebuild a kidney: the lymph node as a possible developmental niche. Cell Transplant. PMID: 26160801

Gill J, Malin M, Hollander GA, Boyd R (2002) Generation of a complete thymic microenvironment by MTS24(+) thymic epithelial cells. Nat Immunol 3(7):635–642

Grompe M (2006) Principles of therapeutic liver repopulation. J Inherit Metab Dis 29 (2–3):421–425

Grompe M, Lindstedt S, al-Dhalimy M, Kennaway NG, Papaconstantinou J, Torres-Ramos CA, Ou CN, Finegold M (1995) Pharmacological correction of neonatal lethal hepatic dysfunction in a murine model of hereditary tyrosinaemia type I. Nat Genet 10(4):453–460

Grompe M, Overturf K, al-Dhalimy M, Finegold M (1998) Therapeutic trials in the murine model of hereditary tyrosinaemia type I: a progress report. J Inherit Metab Dis 21(5):518–531

Gupta S, Vemuru RP, Lee CD, Yerneni PR, Aragona E, Burk RD (1994) Hepatocytes exhibit superior transgene expression after transplantation into liver and spleen compared with peritoneal cavity or dorsal fat pad: implications for hepatic gene therapy. Hum Gene Ther 5 (8):959–967

Henne-Bruns D, Kruger U, Sumpelmann D, Lierse W, Kremer B (1991) Intraperitoneal hepatocyte transplantation: morphological results. Virchows Arch A Pathol Anat Histopathol 419 (1):45–50

Hoppo T, Komori J, Manohar R, Stolz DB, Lagasse E (2011) Rescue of lethal hepatic failure by hepatized lymph nodes in mice. Gastroenterology 140(2):656–666.e652

Komori J, Boone L, DeWard A, Hoppo T, Lagasse E (2012) The mouse lymph node as an ectopic transplantation site for multiple tissues. Nat Biotechnol 30(10):976–983

Lagasse E, Connors H, Al-Dhalimy M, Reitsma M, Dohse M, Osborne L, Wang X, Finegold M, Weissman IL, Grompe M (2000) Purified hematopoietic stem cells can differentiate into hepatocytes in vivo. Nat Med 6(11):1229–1234

Lakkis FG, Arakelov A, Konieczny BT, Inoue Y (2000) Immunologic 'ignorance' of vascularized organ transplants in the absence of secondary lymphoid tissue. Nat Med 6(6):686–688

Lipshutz GS, Busuttil RW (2007) Liver transplantation in those of advancing age: the case for transplantation. Liver Transpl 13(10):1355–1357

Maillard I, Schwarz BA, Sambandam A, Fang T, Shestova O, Xu L, Bhandoola A, Pear WS (2006) Notch-dependent T-lineage commitment occurs at extrathymic sites following bone marrow transplantation. Blood 107(9):3511–3519

Markert ML, Devlin BH, McCarthy EA (2010) Thymus transplantation. Clin Immunol 135 (2):236–246

Overturf K, Al-Dhalimy M, Tanguay R, Brantly M, Ou CN, Finegold M, Grompe M (1996) Hepatocytes corrected by gene therapy are selected in vivo in a murine model of hereditary tyrosinaemia type I. Nat Genet 12(3):266–273

Overturf K, al-Dhalimy M, Ou CN, Finegold M, Grompe M (1997) Serial transplantation reveals the stem-cell-like regenerative potential of adult mouse hepatocytes. Am J Pathol 151 (5):1273–1280

Parent AV, Russ HA, Khan IS, Laflam TN, Metzger TC, Anderson MS, Hebrok M (2013) Generation of functional thymic epithelium from human embryonic stem cells that supports host T cell development. Cell Stem Cell 13:219–229

Perkins JD, Halldorson JB, Bakthavatsalam R, Fix OK, Carithers RL Jr, Reyes JD (2009) Should liver transplantation in patients with model for end-stage liver disease scores <or= 14 be avoided? A decision analysis approach. Liver Transpl 15(2):242–254

Poznansky MC, Evans RH, Foxall RB, Olszak IT, Piascik AH, Hartman KE, Brander C, Meyer TH, Pykett MJ, Chabner KT, Kalams SA, Rosenzweig M, Scadden DT (2000) Efficient generation of human T cells from a tissue-engineered thymic organoid. Nat Biotechnol 18 (7):729–734

Reuben A (2007) Alcohol and the liver. Curr Opin Gastroenterol 23(3):283–291

Rhim JA, Sandgren EP, Degen JL, Palmiter RD, Brinster RL (1994) Replacement of diseased mouse liver by hepatic cell transplantation. Science 263(5150):1149–1152

Rodewald HR (2000) Thymus epithelial cell reaggregate grafts. Curr Top Microbiol Immunol 251:101–108

Seach N, Mattesich M, Abberton K, Matsuda K, Tilkorn DJ, Rophael J, Boyd RL, Morrison WA (2010) Vascularized tissue engineering mouse chamber model supports thymopoiesis of ectopic thymus tissue grafts. Tissue Eng Part C Methods 16(3):543–551

Smith MK, Mooney DJ (2007) Hypoxia leads to necrotic hepatocyte death. J Biomed Mater Res A 80(3):520–529

Starzl TE, Groth CG, Brettschneider L, Penn I, Fulginiti VA, Moon JB, Blanchard H, Martin AJ Jr, Porter KA (1968) Orthotopic homotransplantation of the human liver. Ann Surg 168 (3):392–415

Strom SC, Chowdhury JR, Fox IJ (1999) Hepatocyte transplantation for the treatment of human disease. Semin Liver Dis 19(1):39–48

Terra R, Louis I, Le Blanc R, Ouellet S, Zuniga-Pflucker JC, Perreault C (2005) T-cell generation by lymph node resident progenitor cells. Blood 106(1):193–200

Volk ML, Hernandez JC, Lok AS, Marrero JA (2007) Modified Charlson comorbidity index for predicting survival after liver transplantation. Liver Transpl 13(11):1515–1520

von Andrian UH (1996) Intravital microscopy of the peripheral lymph node microcirculation in mice. Microcirculation 3(3):287–300

Weglarz TC, Degen JL, Sandgren EP (2000) Hepatocyte transplantation into diseased mouse liver. Kinetics of parenchymal repopulation and identification of the proliferative capacity of tetraploid and octaploid hepatocytes. Am J Pathol 157(6):1963–1974

Weir GC (2012) Cellular transplantation into lymph nodes may not be such a crazy idea. Cell Stem Cell 11(5):587–588

Zakrzewski JL, Tuckett AZ, van den Brink MR (2012) Transforming lymph nodes into tissue factories. Nat Biotechnol 30(10):958–959

Chapter 9
Engineering of Artificial Lymph Node

Makoto Nakamura, Kenichi Arai, Takahito Mimura, Jungo Tagawa, Hirotoshi Yoshida, Koichi Kato, Tadashi Nakaji-Hirabayashi, Yuka Kobayashi, and Takeshi Watanabe

Abstract Synthetic immunology has been a topic of research in both the immunology and tissue engineering fields. Since the first successful generation of an artificial lymph node (LN) *in vivo* by implantation of a collagen scaffold with stromal cells has been reported, tissue engineering of artificial LNs using several biological and engineering techniques has started. Recently, the advance engineering approaches using several manufacturing machines, including three-dimensional (3D) printers, 3D fabrication and assembling machines, have been addressed. Such approaches are in general called "Biofabrication". As the use of machines has several potential capabilities, such as super-manufacturing beyond the capability of humans, direct tissue manufacturing by additive manufacturing and computer aided tissue engineering by applying high efficiency of the computer and digital technologies, the strategy of biofabrication can provide a potential of breakthrough to construct highly sophisticated 3D scaffolds and to generate several complex 3D tissues and organs. Based on this concept, we have ever developed several machines for this purpose such as an inkjet 3D bioprinter to construct complex and multi-composite 3D structures. In the present work, we applied this strategy to rationally design and construct artificial LNs. This research is still in its infancy, and it requires several improvements in terms of new technology. Here, we introduced our struggles and current works towards production of practical artificial LNs.

M. Nakamura, M.D., Ph.D. (✉) • K. Arai • T. Mimura • J. Tagawa • H. Yoshida • T. Nakaji-Hirabayashi
Graduate School of Science and Engineering for Research, University of Toyama, Toyama, Japan
e-mail: maknaka@eng.u-toyama.ac.jp

K. Kato
Institute of Biomedical and Health Sciences, Hiroshima University, Higashihiroshima, Japan

Y. Kobayashi • T. Watanabe
Medical Research Institute, Kitano Hospital, Osaka, Japan

T. Watanabe, Y. Takahama (eds.), *Synthetic Immunology*,
DOI 10.1007/978-4-431-56027-2_9

List of Abbreviations

LN	Lymph node
2D	two dimension or two dimensional
3D	three dimension or three dimensional
CAD	computer aided design
CAM	Computer aided manufacturing
CAE	Computer aided engineering
AM	Additive Manufacturing
FDM	fused deposition modeling
UV	Ultra violet
PGA	polyglycolic acid
PLA	polylactic acid
PCL	poly-caprolactone
SLOs	secondary lymphoid organs
TLOs	tertiary lymphoid organs
LTo	lymphoid tissue organizer
LTi	lymphoid tissue inducer
VCAM-1	vascular cell adhesion molecule-1
DC	dendritic cell
FDC	follicular dendritic cell
EC	endothelial cell
SCID mouse	severe combined immunodeficiency mouse
ECM	Extracellular Matrix
MEMS	Micro Electro Mechanical Systems
GF	Growth factor

Overview of Current Situation of Tissue Engineering and Biofabrication

Research Related to Tissue Engineering

In studies of tissue engineering and regenerative medicine, various approaches and technologies have been developed according to the specific tissues being targeted. The most intrinsic purpose of such research is to produce functional cell, tissue, and organ substitutes using biological and bio-engineering procedures, in order to repair, replace, or enhance biological functions that have been lost by disease or injury (Langer and Vacanti 1993). In this context, artificially engineered three-dimensional (3D) tissues have been found promising as effective tools for life science studies including various mechanisms of biological phenomena, pathogenesis, and tissue regeneration (Griffith and Naughton 2002; Khetani and Bhatia 2006). In addition, engineered 3D tissues are promising for effective drug screening and chemical toxicological tests. For developing such applications, various researchers have promoted investigations of tissue engineering to produce various artificial tissues and organs.

Biofabrication: Tissue Engineering Through Development and Use of Machine Technology

In order to overcome several problems in the field of tissue engineering, various approaches using 2D and 3D fabrication machines have been developed, which are generally classified as "bioprinting, biofabrication and bioassembly" (Mironov et al. 2006, 2009; Guillemot et al. 2010; Jakab et al. 2006; Sun 2009). Bioprinting can be defined as the use of computer-aided printing, transfer, and manufacturing processes for patterning, printing and 3D construction with living and non-living biological materials to produce prescribed 2D or 3D tissues and organs. The concepts of biofabrication and bioassembly are, strictly speaking, somewhat wider than that of bioprinting, because they include not only 2D and 3D printing techniques but also all available manufacturing and assembling technologies, such as rolling, lapping, knitting, laminating, assembling, building, and others. However, all of them are aimed to computerize, digitize, and mechanize various tissue engineering processes for the final goal of producing living and functional organs. In this paper, we include these 3 approaches in the term biofabrication (Fig. 9.1).

The general concept of biofabrication, shown in Fig. 9.1, was originally based on application of computer aided design (CAD), manufacturing (CAM), and engineering (CAE) techniques initially employed in industrial manufacturing to the tissue engineering field (Sun and Lal 2004; Mironov et al. 2003). CAD/CAM/CAE approaches have brought dramatic innovations to the manufacturing industry by introducing computer technology. Bio-CAD and Bio-CAM represent computer aided design and manufacturing for biofabrication, respectively. Products produced

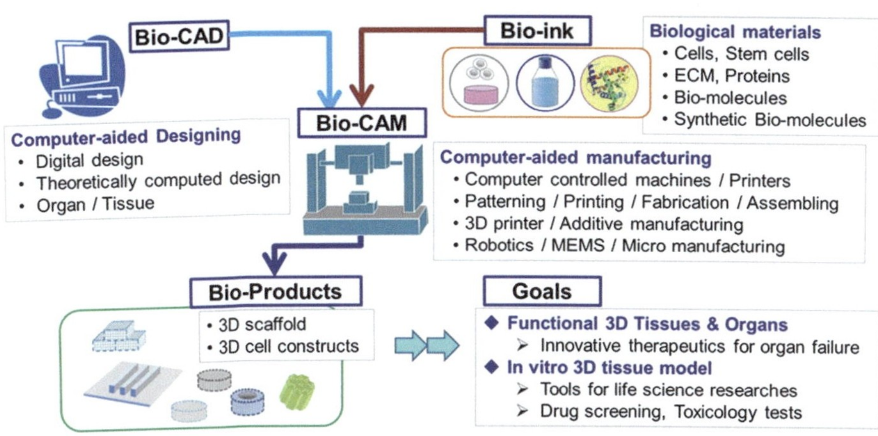

Fig. 9.1 General concepts of bioprinting/biofabrication/bioassembling

Table 9.1 Advantages of introduction of manufacturing machines to tissue engineering

Advantage of the use of machines	Advantages in tissue engineering
1) Super high quality manufacturing by machines	
• High precision manufacturing/ high resolution	• Manufacturing of bio-products with microscopic structures
• High speed manufacturing / High productivity	• Manipulation of huge numbers of cells
• High reproducibility	• Manufacturing reproductive bio-products
2) Direct manipulation of constituent materials	
• Direct mounting of cells and biological • materials Additive manufacturing	• On demand direct tissue engineering • Bottom-up tissue engineering
3) Availability of high performance computer	
• Computer aided designing/ manufacturing/ engineering	• Computer aided tissue engineering CAD /CAM/CAE in tissue engineering

using a Bio-CAM process are termed Bio-products, while Bio-ink represents the materials used for the composition of bio-products. Biological materials such as cells, extracellular matrices and bio-functional molecules are important natural Bio-ink, while synthetic materials and the materials used in artificial organs are also possible artificial Bio-inks. Furthermore, Bio-ink can be defined as a material for Bio-CAM. As ink materials have contributed to the development of high quality printing, the development of Bio-ink is also one of the important keys for high quality Bio-CAM, especially for obtaining high quality tissue and organ regeneration, which are essential for fully designed and engineered bio-artificial organs.

One of the important characteristics of biofabrication is the use of computer controlled manufacturing machines or printers for Bio-CAM. The major advantages of the use of manufacturing machines for tissue engineering are summarized in Table 9.1. For example, super precision and super high-speed manufacturing beyond human capability will become possible, and novel tissue manufacturing approaches can be realized by developing and applying innovative machines.

Construction of 3D structures including internal structures has been an important issue in tissue engineering. In conventional tissue engineering, 3D scaffolds are used, many of which are produced by a molding method, in which a 3D mold is used and filled with material, or by subtraction manufacturing, in which a 3D solid mass is shaved with a cuting apparatus or drill. However, it is impossible to construct internal structures using these approaches. In addition, as cells are seeded onto a 3D scaffold after it is manufactured, cells can be seeded only on the surface, not inside of the scaffold. On the other hand, the structures and the composition of the interior of the tissues are indispensable for many biologically functional tissues and organs, including the arrangement and the composition of cells. For this reason, the technology which achieves that has been wished for. Then, the bottom-up approach, which includes a direct manufacturing and a layered laminating printing approach, can meet this requirement. Such a bottom-up approach is called additive manufacturing (AM). The constituent materials are positioned and mounted

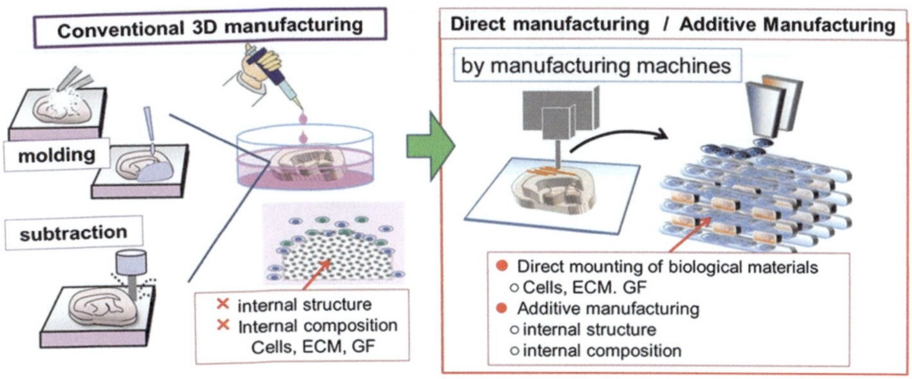

Fig. 9.2 Schema of the additive manufacturing based direct tissue engineering

according to a prescript design in a direct manufacturing, while they are stacked in 3D spatial positions through a layered laminating printing approach. This AM based direct 3D tissue engineering is shown schematically in Fig. 9.2, as compared to conventional 3D tissue engineering. Presently, AM is considered to be the most suitable and promising approach for biofabrication and 3D tissue engineering.

CAD/CAM/CAE based machines have great potential for tissue engineering, thus we have proposed bioprinting and biofabrication researches, and endeavored to develop some promising machines and technologies.

Currently Available 3D Printers for AM

Since its invention in 1980 in Japan, 3D printing technology has continued to be developed. In the State of the Union address in February 2013, United States President Barack Obama noted that "3D printing has the potential to revolutionize the way we make almost everything". Following that remark, 3D printing as well as AM have become the center of attention throughout the world, and various types of 3D printers have since been developed. And several Bio-3D printers have also been developed for medical and biological fields.

Figure 9.3 shows a classification of current industrial 3D printers, including those in a recent Japanese review (Nakamura et al. 2012). In general, industrial 3D printers are classified into 5 major types; stereo lithography, fused deposition modeling (FDM), particle deposition, powder based 3D printing, and sheet lamination.

Stereo lithography has the longest history and remains as a major 3D printing procedure (Fig. 9.3a). With this type, a focused UV laser beam solidifies UV curative resin in layers and 3D structures can be constructed in a bath of UV curable resin. Furthermore, 3D stereo lithography that uses a UV projector has also been developed. Although color printing is difficult, high resolution 3D fabrication is

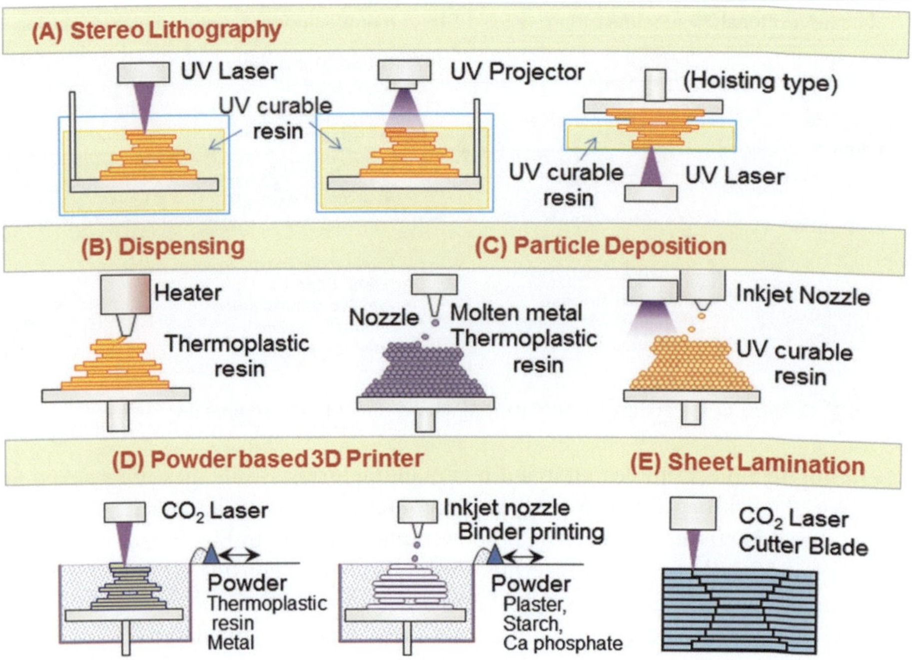

Fig. 9.3 Current industrial 3D printers

possible by applying 2-photon polymerization. FDM is presently the most popular 3D printing method (Fig. 9.3b). With this, ink materials are extruded from nozzles by air pressure or syringe extrusion onto a computer controlled XYZ stage. Several scaffolds for tissue engineering have been constructed with thermoplastic biodegradable materials and tested. Although the resolution is limited, the mechanism is quite simple. The particle deposition method is an approach in which particles of ink material are ejected from a nozzle and deposited in layers while solidifying the molten ink materials by temperature change (Fig. 9.3c). A UV inkjet 3D printer using UV curable ink materials is one of the advanced models of this type. An inkjet system can also be used as a method to deposit ink, after which the deposited ink droplets are solidified by UV irradiation. Powder based 3D printers are divided into two types (Fig. 9.3d), laser sintering and inkjet. Solidification of powder particles is caused by melting due to focused laser heating in the former, while inkjet glue is used for printing in the latter. After writing patterns, the fabrication platform is formed in a single step and covered again with powders. These processes are repeated until the whole of the 3D structure is constructed in the powder. The sheet lamination method is an approach to construct 3D structures by laminating paper sheets in sequence (Fig. 9.3e). High resolution color 3D printing is realized by the use of inkjet printing. Each laminated paper sheet is cut in an arbitrary manner by a laser beam or fine cutter blade according to each sectional image of the 3D shape. Finally, a 3D paper structure can be obtained.

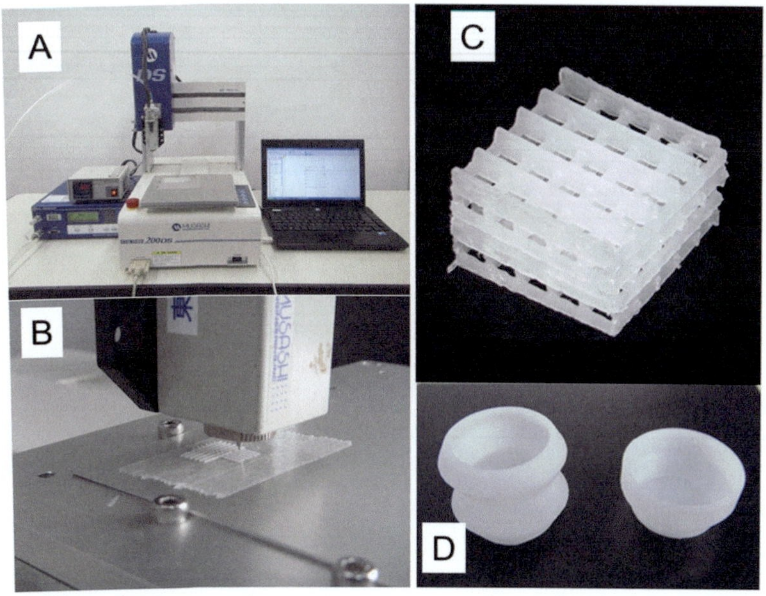

Fig. 9.4 FDM type 3D printers and fabrication of 3D scaffolds. PCL was used as a Bio-ink of biodegradable thermoplastic polymer (Photographs were offered by Musashi Engineering Ink, Tokyo, Japan, http://www.musashi-engineering.co.jp/)

Each type of 3D printer has both strong and weak points. Their application to tissue engineering was started from fabrication of 3D scaffolds using biocompatible and biodegradable materials, for example, powder-based 3D printing with hydroxy-apatite powder and FDM with thermoplastic polymers, such as polyglycolic acid (PGA), polylactic acid (PLA), and poly-caprolactone (PCL) (Vozzi et al. 2002; Hutmacher et al. 2004; Leong et al. 2008). These are generally used for fabrication of bone and cartilage. As the performance of 3D printer devices has advanced, the quality of 3D scaffolds has also improved. Figure 9.4 shows a representative product recently produced by FDM-based 3D printing.

However, their use is limited to fabrication of acellular scaffolds, which can be used only for conventional scaffold based tissue engineering and not direct tissue manufacturing. As a result, special Bio-3D printers have been developed in order to directly utilize biological materials including living cells, proteins, and bio-functional molecules. Current Bio-3D printers are reviewed and summarized in Fig. 9.5 (Bajaj et al. 2014). In order to keep living cells alive, biocompatible hydrogels are used for 3D structures, such as fibrin, alginate hydrogels, and synthetic hydrogels. Cell spheroids are also used instead of individual cells in a 3D Bio-plotter using a dispenser technique (Fig. 9.5-II-B), as adjacent spheroids make a connection and fuse to regenerate tissue structures (Jakab et al. 2004; Faulkner-Jones et al. 2013). Scaffold free tissue engineering has been proposed using this type, however, some structures to support 3D assemblies temporarily just like a scaffold are needed in practice, such as mold or micro-needles.

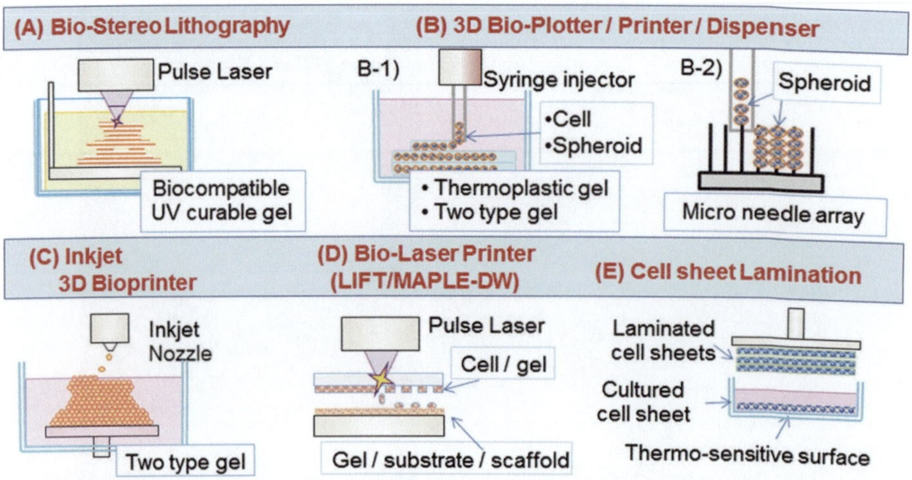

Fig. 9.5 Current Bio-3D printers

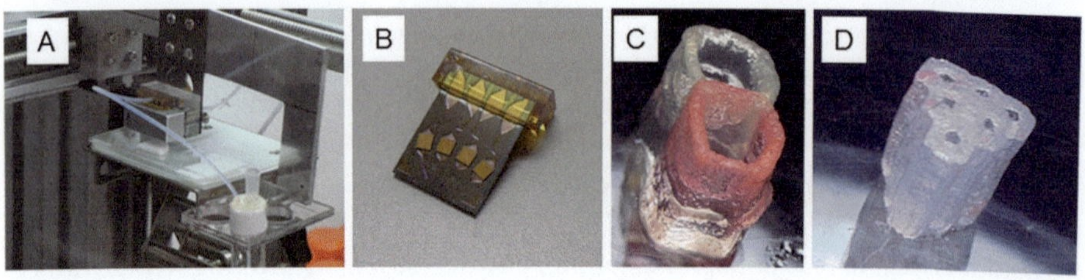

Fig. 9.6 Developed inkjet 3D bioprinter for Bio-CAM. Printer device (**a**), piezo type inkjet head (**b**), produced 3D hydrogel structures (**c**, **d**)

We developed the inkjet 3D bioprinter shown in Figs. 9.5c and 9.6 (Nakamura et al. 2005; Nishiyama et al. 2009; Nakamura 2010, 2013). Inkjet printing technology has great potential for precision manufacturing with high resolution, high speed manipulation of individual cells, and multi-color printing with different materials, all of which are thought the essential requirements for 3D biofabrication of complex tissues and organs. Two type gels, which are formed by mixing of gel precursor and gel reactor solution, can be used, such as alginate and fibrin gel. However, biologically raw materials limit the fabrication procedures and process conditions, such as mechanical force, temperature, biocompatibility of the materials used, and cell environment. In addition, construction of micro- to macro-sized structures, control tissue regeneration, and vascular induction are the next important issues to be dealt with. Bio-3D printers for direct manufacturing as well as 3D tissue and organ engineering are still in the development stage, with several issues remaining to be overcome.

Background of Synthetic Immunology

Research of Immune System

The immune system is an important physiological system to protect from a variety of infectious foreign bodies such as viruses and bacteria. Its mechanism is extremely complicated and it functions in a dynamic manner. A wide variety of foreign bodies must be recognized and memorized, and a specific immune response against such foreign body must be generated. Such immune reactions are sometimes limited within a local area and they sometimes spread systemically throughout the body. Usually, they are within physiologic limits, however, sometimes they bring pathogenic phenomena to the host as well. Immune reactions to foreign invaders must be appropriately regulated in all phases of generation, maintenance, termination, restoration, and reactivation.

Because of the importance and the complexity of the immune system, immunology has been explored with much interest. Various types of related immune cells have been distinguished as important players in immune functions and their respective roles have also been elucidated. Furthermore, a variety of signal molecules, such as cytokines, chemokines, and growth factors, as well as their respective surface markers and receptor proteins have also been discovered. Procedures and various instrumentations needed to recognize and analyze such molecules and cell types have been developed, including many types of fluorescent antibodies, and fluorescent stain and flow cytometry systems. Owing to those developments, detailed findings about the immune system have been presented.

Synthetic Immunology

Based on the background described above, engineering of immunological tissues has become an emerging trend in the fields of immunology and tissue engineering, and related research is referred to as synthetic immunology. This can be defined as the reconstruction of bio-artificial immune systems based on rational design to perform complex immunological functions (Spiegel 2010; Geering and Fussenegger 2015).

Artificially reconstructed immune systems are expected to contribute to at least two important advancements. The first is to provide effective tools for research of complex immunological phenomena and mechanisms. Especially the wide variety of interactions between cell and cell, cell and the ambient environment in the spatial field of immunity can be investigated effectively with spatial designed artificial immune systems. The second is for development of innovative clinical therapies to treat various diseases, such as intractable infectious diseases, autoimmune diseases, and cancer, using artificial immune systems as a therapeutic device which can control immune reactions. For both, accumulated knowledge regarding immune mechanisms and sufficient preparation of various tools to recognize immune reactions, immune cells and immune molecules are extremely advantageous. With those in hand, research of artificial immune systems is very promising. Therefore, synthetic immunology can serve to take immunology and tissue engineering to the next stage.

Artificial LNs

Artificial LNs Development as a Major Target of Synthetic Immunology

LNs are important immunological organs that serve to provide major adaptive immune responses. Adaptive immunity, also called acquired immunity, is essential for the progression of immunity in humans and other mammals. It is known that adaptive immune responses are supported by secondary lymphoid organs (SLOs), such as LNs, Peyer's patches and the spleen, and tertiary lymphoid organs (TLOs), which often develop at sites of chronic inflammation. Both types of organs serve to recognize invading antigens and present them to lymphocytes, then humoral and cellular immunities are generated and function. Because of this acquired immune system, antigen-specific immunity can be acquired against millions of foreign antigens. The LNs are the most fundamental organ type involved in this process, thus the ability to manufacture artificial LNs is a very important theme for synthetic immunology.

Previous Research of Artificial LNs

Watanabe et al. attempted to engineer artificial LNs and reviewed their attempts for synthesis of artificial lymphoid tissues (Suematsu and Watanabe 2004; Kobayashi and Watanabe 2010; Cupedo et al. 2012). In those studies, related cells, soluble factors, and their combinations were also reviewed and summarized as basic information. They focused on the importance of stromal cells as lymphoid tissue organizer (LTo) cells. In addition, successful *in vivo* regeneration was also reported, analyzed, and discussed.

During embryogenesis, it is known that stromal LTo cells attract and are activated by circulating lymphoid tissue inducer (LTi) cells to express the adhesion molecule vascular cell adhesion molecule-1 (VCAM-1) and secrete the lymphoid chemokines CXCL13, CXCL12, and CCL19/CCL21. VCAM1 is a cell adhesive molecule expressed on the cell membrane of vascular endothelial cells, macrophages, and stromal cells, as well as others, and mediates the adhesion of white blood cells including lymphocytes onto other cells expressing VCAM-1. CXCL13 is known to initiate lymphoid organ formation (van de Pavert et al. 2009; Ansel et al. 2000) and migration of B cells into follicles, and also promotes follicular dendritic cell (FDC) development (Okamoto et al. 2007). CXCL12 contributes to T cell trafficking into LNs, while CCL19 and CCL21 attract and retain T cells, respectively. These functions of LTo stromal cells are shown in Fig. 9.7.

Figure 9.8 presents the schema of their strategy for producing artificial LNs. Their first successful *in vivo* regeneration of LN-like tissue was accomplished by implantation of a collagen sponge together with stromal cells and bone marrow-derived dendritic cells (DCs), as shown in Fig. 9.8A. After transplantion into renal subcapsular space of experimental mice, grafts formed structures the same as SLOs (Suematsu and Watanabe 2004). Strong secondary immune responses after re-immunization were also examined and confirmed (Okamoto et al. 2007). In addition, these artificial LNs were confirmed to function after transplantation into SCID mice.

Their next challenge was generation of artificial LNs without using stromal cells, based on the fact that stromal cells induce secondary immune reactions by secreting several lymphoid chemokines and inducing cell adhesion molecules. Collagen

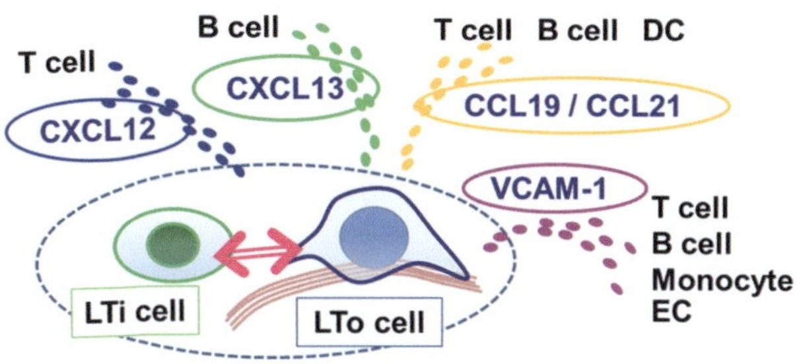

Fig. 9.7 Schema of the principle of secondary lymphatic organ (SLO) formation. *LTo cells* lymphoid tissue organizer stromal cells, *LTi cells* lymphoid tissue inducer cells, *DC* dendlic cells, *EC* endothelial cells of capillary vessels and lymphoid venules

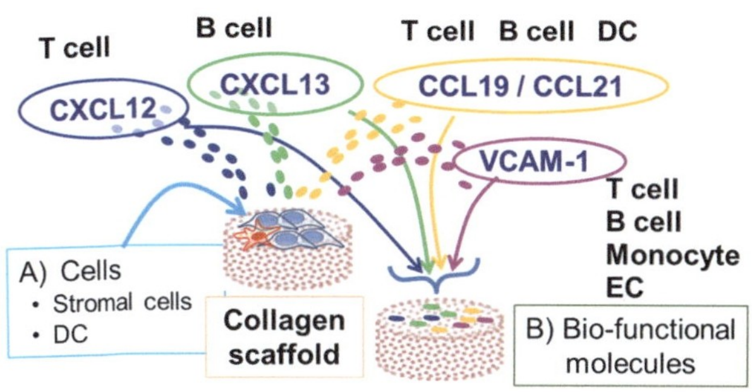

Fig. 9.8 Strategy of artificial lymph node. Collagen scaffold with stromal cells and DC (**A**) and with bio-functional molecules (**B**)

sponges containing lymphoid chemokines and adhesion molecules were prepared, and then implanted into mice, instead of collagen sponges with stromal cells (Fig. 9.8b). After 3 weeks, formations of B and T cell clusters, and vascular system invasion were observed in mice with the implanted grafts. Therefore, it was concluded that those chemokines play important roles for *in situ* lymphoid tissue regeneration. Synthesized bio-functional molecules were used in that study in order to prolong their functions (see next section) and lymph node-like structures were successfully shown by histological findings following that procedure. However, secondary immune responses induced in the grafts were not sufficient. Those results suggested that synthetic chemokines are significantly effective for attraction and accumulation of immune cells, while the existence of stromal LTo and LTi cells remains essential for complete immune functions. Pursuit of greater details regarding cell-cell and cell-chemokine interactions, as well as their spatial interactions are very interesting issues in immunology, and this approach is considered to be a promising method for practical development of artificial LNs without using living cells.

Synthesized Chemokines

Natural chemokines are usually water soluble and soon diffused after immersion in culture medium or implanted *in vivo*. On the other hand, a few days to a few weeks is required to form LN structures by accumulation of native circulating B and T lymphocytes. Therefore, the functions of soluble chemokines must be maintained for a certain period until significant formation of LN structures. For this purpose, recombinant bio-functional molecules connected with collagen binding peptides were designed and prepared. Owing to collaboration among biomaterial researchers, 4 types of synthesized chemokines, CXCL13, CXCL12, CCL19/CCL21, and VCAM1, were prepared (Fig. 9.9). These are anticipated to be fixed onto collagen molecules by biological interactions between the collagen binding domain of the synthetic molecules and respective peptide domain of the collagen

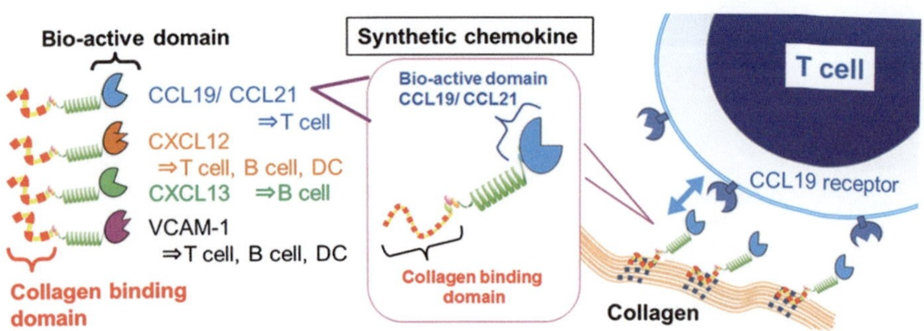

Fig. 9.9 Synthetic chemokines and adhesive molecule

molecules. As a result, it is expected that the functions of these chemokines will be maintained for a longer period in place where they are bound.

Biofabrication of Artificial LNs

Strategy for Development of Synthetic LNs (Fig. 9.10)

Following is a summary of advances needed for development of artificial LNs and related important issues to be solved. (1) Accumulation of biological and medical information. Understanding of immunology, histology, and molecular cell biology, as well as interactions between cells, and cells and chemokines, as well as their spatial relationship is essential for designing theoretically rationale artificial LNs. (2) Selection of suitable biological materials for cells, biomaterials, and bio-functional molecules. Suitable synthetic materials or molecules should also be developed as necessary. (3) Tissue engineering. Based on Bio-CAD and Bio-CAM and the use of Bio-ink, theoretically suitable artificial LNs should be produced as a Bio-product. High quality design and construction using computer and manufacturing technologies are expected to produce excellent artificial LNs. (4) Production of artificial LNs. For artificial LNs, 3D scaffolds with cells and those with chemokines without cells have been developed and shown to be potentially effective. (5) Implantation *in vivo*. Kidney subcapsular space has been successfully used for implantation. (6) *In situ* regeneration. Implanted artificial LNs attract lymphoid cells and finally SLOs with the characteristics of histological structures are regenerated *in vivo*. *In vitro* regeneration of SLOs has also attracted attention and is an important research theme, though we will not provide additional details regarding that in this paper.

After investigating synthetic immunology, our focus turned to fabrication of artificial LNs by application and integration of biofabrication technology. Next, we show our approach for tissue engineering, especially biofabrication procedures.

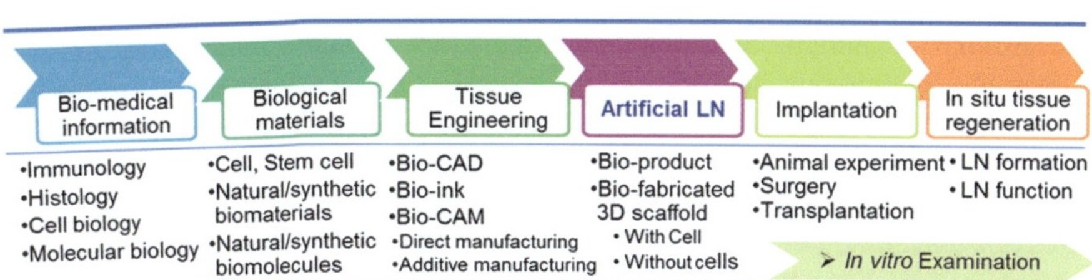

Fig. 9.10 Outline of the engineering of artificial lymph node

Designing Artificial LNs Based on Immunological Background

Artificial LNs with and Without Cells

It has been suggested that the functions of stromal cells are essential, thus production of artificial LNs containing living cells is reasonable. However, from a practical viewpoint, a scaffold with chemokines without cells is thought to be more realistic for artificial organs, primarily because long-term preservation is difficult and several cell culture processes are required when living cells are used.

Artificial LNs with Chemokines

The majority of cells that compose a lymph node, such as T and B lymphocytes, are generated in bone marrow. They then circulate throughout the body and enter into lymph nodes via afferent lymphatic vessels, and exit via efferent lymphatic vessels. They accumulate in LNs and finally SLOs are generated.

For an effective artificial LN design, it is necessary to take the process and time course of LN regeneration into account. In addition, artificial LNs should have a function to attract circulating immune cells, for which chemokines and adhesive molecules are essential key factors. An important function of a scaffold for producing artificial LNs is the potential to provide space where immune cells can adhere, accumulate, proliferate, and communicate with other to exchange immunological information.

As noted above, Watanabe and colleagues focused on CXCL13, CXCL12, and CCL19/CCL21 as important chemokines, as well as VCAM1 as an essential adhesive molecule. Based on their immunological understanding, they prepared respective synthetic molecules and showed that they were effective for morphogenesis of LNs *in vivo*. The chemokines CXCL13, CXCL12, CCL19, and CCL21 correspond to B, T, and B cells, and DCs, respectively, while the cell adhesive molecule VCAM1 corresponds to all of those cells as well as vascular related cells.

The LN geometrical structure has already been shown in histological findings. Among several histological structures existing in LNs, lymphatic nodules, in which several types of immune cells are accumulated, are especially important for their immunological function. In addition, there are some localized domain structures, such as follicles composed of mainly B cells and follicle-neighborhood domains where T cells are dominant, as shown in Fig. 9.11.

When considering the above information in regard to the histology of LNs, origins of constituent cells, the relationship between those cells and chemokines, and previous findings of synthetic LNs, we expect that a 3D scaffold design in which cell-specific chemokines are geometrically positioned and fixed based on histological patterns will be the most practical for artificial LN development. For

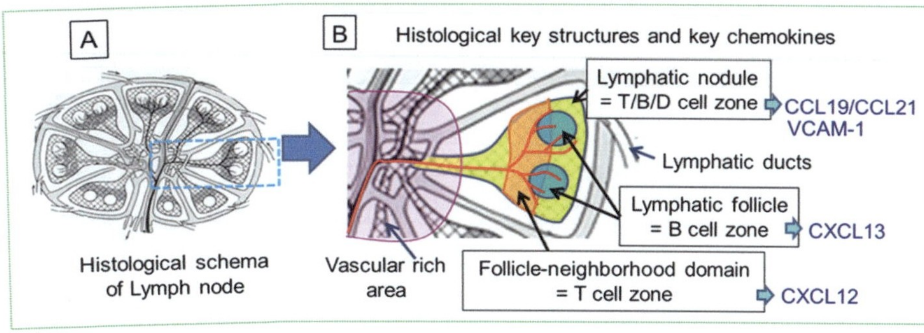

Fig. 9.11 Rational design of artificial LN. Histological schema of lymph node (**a**), rational design of artificial LN based on the histological key structures and key chemokines (**b**). Locations of the key chemokines and key adhesive molecule

fabrication of such artificial LNs, an additive manufacturing approach with 3D bioprinter is thought to be effective.

Biofabrication Using Inkjet 3D Bioprinter

As noted above and shown in Fig. 9.6, we developed an original custom-made inkjet 3D bioprinter that shows promise as a Bio-3D printer. In addition, we have been able to prepare synthetic collagen binding chemokines. Thus, fabrication of an effective 3D scaffold using collagen is our next targeted issue.

For our inkjet 3D bioprinter, two types of hydrogel are needed, though alginate hydrogel, produced by mixing two isolated solutions of sodium alginate and calcium chloride, is mainly used. The gel reaction is based on ionic affinity and quite rapid, thus it is advantageous for biofabrication. On the other hand, time is required to form collagen gel for 3D fabrication. In addition, the viscosity of collagen is too high for inkjet printing, thus it is difficult to use without modification.

We focused on collagen peptide, which is a degradation product made by hydrolyzing gelatin. Its molecular weight is below 50,000, while that of collagen and gelatin is 300,000 and 100,000, respectively. Although collagen peptide is water soluble and has a much lower viscosity, it cannot form gel. Thus, we attempted to modify alginate molecules with collagen peptide through chemical modification. The alginate molecule has carboxyl moieties, while collagen peptide has amino moieties. Based on a report by Kamimura et al. (Kamimura et al. 2012), we synthesized alginate molecules modified with collagen peptide through dehydration-condensation reactions (Fig. 9.12) and successfully obtained collagen peptide-modified alginate. It was confirmed to have an ability to form gel with calcium ion just like sodium alginate, and the viscosity was significantly decreased and useable in a piezoelectric inkjet 3D bioprinter. We have also confirmed that the

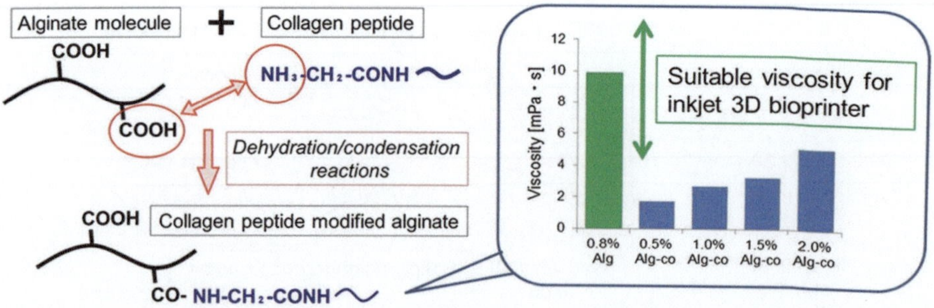

Fig. 9.12 Development of collagen modified alginate molecules for inkjet 3D bioprinting. *Left*; synthesis of the collagen peptide modified alginate molecular, *Right*; the adjustment of the viscosity of the collagen peptide modified alginate to be used

abilities of cell adhesion and cell spreading are superior as compared to non-modified alginate gel. We are now examining the effectiveness of synthesized alginate gel for artificial LNs.

Biofabrication by Extrusion Using Micro-Manufactured Spinneret Nozzles

Application of Spinneret Nozzle

Collagen is an essential natural biomaterial for biofabrication, though too viscous to be ejected by an inkjet, for which synthesized collagen peptide-modified alginate, as noted above, is a possible solution. We have also tried another approach instead of an inkjet.

We investigated the feasibility of micro-manufactured spinneret nozzles for production of special composite fibers with fine 3D structures with multi-composite materials. In the textile industry, various spinneret nozzles, the orifice part of the spinning device through which melted polymer is extruded as fibers or threads, have been manufactured and used. Various spinneret nozzles are available, such as a special micro-shaped orifice and nozzles with special interior structures that form a complex flow path. Those with a special micro-shaped orifice can produce fine fibers with unique shapes in a cross-section, while nozzles with unique interior structures can produce multi-composite fibers containing several micro-nano-component fibers composed of different materials. Therefore, application of micro-fabricated spinneret nozzles is thought to be promising for producing complicated composite fibers with collagen.

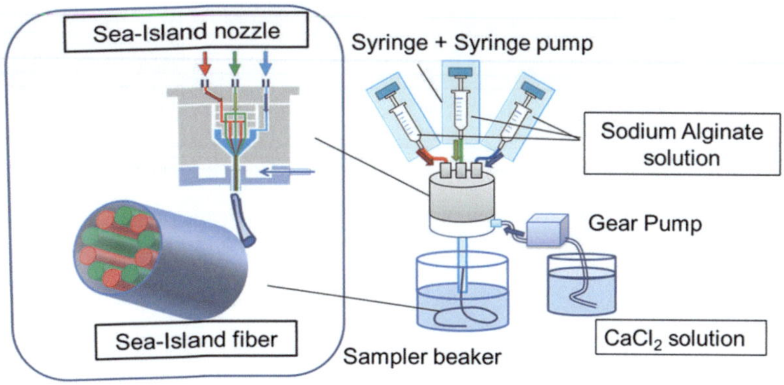

Fig. 9.13 Experimental set up for producing multi-composite fiber using a Sea-Island spinneret nozzle

Experimental Design Using Micro-Fabricated Spinneret Nozzles

We used Sea-Island nozzles (Kasen Nozzle Mfg. Co. Ltd., Osaka, Japan), equipped with three different multi-material flow channels to produce multi-composite fibers, in collaboration with the manufacturer. The schema of the preliminary experiment is shown in Fig. 9.13. The nozzle parts were primarily made of stainless steel or acrylate, and they were precisely processed with microfabrication technology especially for this study. With this nozzle, two different materials can be infused from two different inlet ports and those respective flows are then divided into five different flow passes, then alternatively through 10 island extrusion ports arranged in a circle, which is for the production of 10 Island parts. From the third inlet port, a third material can be infused to produce the Sea part. Finally, Sea-Island fibers with 10 islands can be produced in a sample beaker.

Each inlet port was connected to a syringe containing one of the materials and syringe pumps were used to extrude the gel precursor materials. In preliminary experiments, we used several concentrations of sodium alginate solution prepared mixed with different fluorescent color marker inks for visualization. In this manner, a non-stained Sea fiber, in which green and red fluorescent Island fibers were included, was produced. A 2 % $CaCl_2$ solution was used as the gelation reacting agent and infused from the fourth inlet port by a gear pump. Solution from this inlet port flows at the most outside portion of the extruded fibers and solidifies them into multi-composite gel fibers. The outlet port was guided by a silicon tube to a large sized beaker to hold the sample and the nozzle was assembled in water in order to deaerate it. Thus, all flow passing through the device was filled with water. Before the actual experiments, the water inside the nozzle was replaced with an alginate-Na solution. The flow rate and concentration of the alginate-Na solution were changed based on trial and error. To investigate various shapes and structures of the produced fibers, those were embedded in 5 % agar gel, then cross-sections

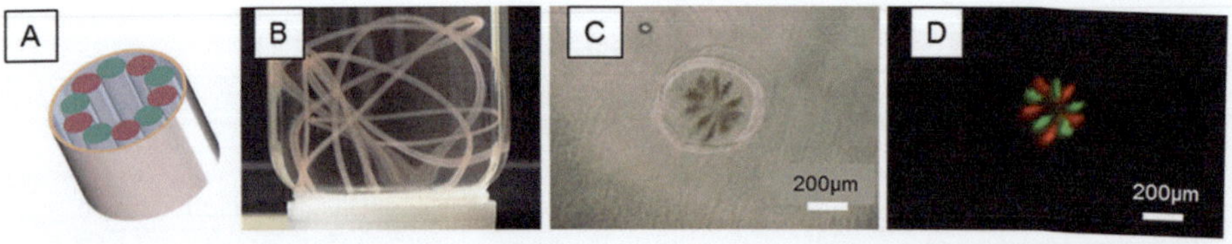

Fig. 9.14 Produced multi-composite fiber of alginate hydrogel. Schema of the composite fiber (**a**), photograph of produced fiber (**b**), cross sectional image of phase contrast image (**c**) and fluorescent image (**d**)

were made and observed using a phase contrast microscope and fluorescence microscope.

Results of Experiments With Sea-Island Nozzle

Representative fibers produced with the Sea-Island nozzle are shown in Fig. 9.14. It was observed that the 10 Island parts with different two colored alginate-Na solutions were arranged alternatively in a round formation in the non-stained alginate fibers. Although the conditions of production must be further explored until undistorted fibers can be produced composite fibers in which multiple materials were arranged were successfully produced using Sea-Island spinneret nozzles.

Conclusion and Future Perspectives for Engineering of Artificial LNs

In this paper, we introduced important concepts and the current status of biofabrication technology with 3D manufacturing machines with a view towards successful production of artificial LNs by engineering. High performance machines will enable AM to realize direct tissue manufacturing, while advantages in regard to tissue engineering are also discussed. Owing to collaboration with researchers in the field of synthetic immunology, designing and fabrication of reasonable and practical artificial LNs will be possible based on advanced understanding of biological and immunological processes. With such an approach, suitable manufacturing methods are now being examined, and effective materials and devices are also being developed. In addition, we presented some challenges encountered when using an inkjet 3D bioprinter and application of spinneret nozzles. Through abundant trial and error, we have been able to accomplish significant steps and found the feasibility of both approaches. Although many problems and issues remain to be overcome, we believe that continued steady steps will allow us to achieve the final goal of producing practical artificial LNs, which will contribute to several important factors related to life science and clinical medicine.

Artificial LNs produced through biofabrication technology will also show advantages in regard to practical applications, as they are expected to be of the same quality, and feature the same structures and functions. High reproducibility is another important advantage of use of machines. Since high quality products are required, especially for medical use, bio-products produced by biofabrication with high performance machines show great promise. Biofabrication must be considered as indispensable technology in order to deliver bio-products with high quality to on-site researchers and clinicians.

In conclusion, the age of synthetic immunology has begun. Close collaboration among researches in the fields of immunology and biofabrication technology will lead to pioneering advances in this new age.

Acknowledgements The authors would like to thank the members of the artificial LN project team at Nakamura Laboratory at the University of Toyama and the staff of Kasen Nozzle Mfg. Co. Ltd. for help with the experiments using the spinneret nozzle. We would also like to thank the members of the Grant-in-Aid for Scientific Research on Innovative Areas "Analysis and Synthesis of Multidimensional Immune Organ Network" for important information regarding synthetic immunology. In addition, we express our thanks to Musashi Engineering Inc. and Leica Microsystems K.K. for kindly offering photographs and their kind support for obtaining images. For Fig. 9.9, we are thankful to Professor Shiro Mizoguchi a honorary professor of Kobe University for giving permission of use a figure of the data base of histology of Kobe Gakuin University.

This work was supported by a Grant-in-Aid for Scientific Research on Innovative Areas "Hyper Bio Assembler for 3D Cellular Innovation" (no. 26106713) from the Ministry of Education, Culture, Sports, Science and Technology of Japan (MEXT), and A-STEP (Adaptable & Seamless Technology Transfer Program through Target-driven R&D) from the Japan Science and Technology Agency (no. AS242170P).

References

Ansel KM, Ngo VN, Hyman PL, Luther SA, Förster R, Sedgwick JD, Browning JL, Lipp M, Cyster JG (2000) A chemokine-driven positive feedback loop organizes lymphoid follicles. Nature 406(6793):309–314

Bajaj P, Schweller RM, Khademhosseini A, West JL, Bashir R (2014) 3D biofabrication strategies for tissue engineering and regenerative medicine. Annu Rev Biomed Eng 16:247–276

Cupedo T, Stroock A, Coles M (2012) Application of tissue engineering to the immune system: development of artificial lymph nodes. Front Immunol 3:343

Faulkner-Jones A, Greenhough S, King JA, Gardner J, Courtney A, Shu W (2013) Development of a valve-based cell printer for the formation of human embryonic stem cell spheroid aggregates. Biofabrication 5(1):015013

Geering B, Fussenegger M (2015) Synthetic immunology: modulating the human immune system. Trends Biotechnol 33(2):65–79

Griffith LG, Naughton G (2002) Tissue engineering-current challenges and expanding opportunities. Science 295(5557):1009–1014

Guillemot F, Mironov V, Nakamura M (2010) Bioprinting is coming of age: report from the International Conference on Bioprinting and Biofabrication in Bordeaux (3B'09). Biofabrication 2(1):010201–010207

Hutmacher DW, Sittinger M, Risbud MV (2004) Scaffold-based tissue engineering: rationale for computer-aided design and solid free-form fabrication systems. Trends Biotechnol 7:354–362

Jakab K, Neagu A, Mironov V, Markwald RR, Forgacs G (2004) Engineering biological structures of prescribed shape using self-assembling multicellular systems. PNAS 101:2864–2869
Jakab K, Damon B, Neagu A, Kachurin A, Forgacs G (2006) Three-dimensional tissue constructs built by bioprinting. Biorheology 43:509–513
Kamimura W, Hattori R, Koyama H, Miyata T, Takato T (2012) A calcium-cross-linked hydrogel based on alginate-modified atelocollagen functions as a scaffold material. J Biomater Sci Polym Ed 23(5):609–628
Khetani SR, Bhatia SN (2006) Engineering tissues for in vitro applications. Curr Opin Biotechnol 17(5):524–531
Kobayashi Y, Watanabe T (2010) Synthesis of artificial lymphoid tissue with immunological function. Trends Immunol 31(11):422–428
Langer R, Vacanti JP (1993) Tissue engineering. Science 260(5110):920–926
Leong KF, Chuaa CK, Sudarmadjia N, Yeonga WY (2008) Engineering functionally graded tissue engineering scaffolds. J Mech Behav Biomed Mater 1(2):140–152
Mironov V, Boland T, Trusk T, Forgacs G, Markward RR (2003) Organ printing: computer-aided jet-based 3D tissue engineering. Trends Biotechnol 21(4):157–161
Mironov V, Reis N, Derby B (2006) Bioprinting: a beginning. Tissue Eng 12(4):631–634
Mironov V, Trusk T, Kasyanov V, Little S, Swaja R, Markwald R (2009) Biofabrication: a 21st century manufacturing paradigm. Biofabrication 1(2):022001
Nakamura M (2010) Reconstruction of biological three-dimensional tissues: bioprinting and biofabrication using inkjet technology. In: Ringersen BR, Spargo BJ, Wu P (eds) Part II: inkjet approaches, cell and organ printing. Springer, Heidelberg, pp 23–34
Nakamura M (2013) Tissue engineering, a case study. In: Hutchings IM, Martin GD (eds) Inkjet technology for digital fabrication. Wiley, West Sussex, pp 307–324
Nakamura M, Kobayashi A, Takagi F, Watanabe A, Hiruma Y, Ohuchi K, Iwasaki Y, Horie M, Morita I, Takatani S (2005) Biocompatible inkjet printing technique for designed seeding of individual living cells. Tissue Eng 11:1658–1666
Nakamura M, Arai K, Toda H (2012) J Soc Instrum Control Eng 51(10):907–911. (Japanese)
Nishiyama Y, Nakamura M, Henmi C, Yamaguchi K, Mochizuki S, Nakagawa H, Takiura K (2009) Development of three-dimensional bio-printer: construction of cell supporting structures using hydrogel and state-of-the-art inkjet technology. J Biomech Eng 131(3):035001
Okamoto N, Chihara R, Shimizu C, Nishimoto S, Watanabe T (2007) Artificial lymph nodes induce potent secondary immune responses in naive and immunodeficient mice. J Clin Invest 17(4):997–1007
Spiegel DA (2010) Grand challenge commentary: synthetic immunology to engineer human immunity. Nat Chem Biol 6(12):871–872
Suematsu S, Watanabe T (2004) Generation of a synthetic lymphoid tissue–like organoid in mice. Nat Biotechnol 22(12):1539–1545
Sun W (2009) Welcome to biofabrication. Biofabrication 1(1):010201
Sun W, Lal P (2004) Recent development on computer aided tissue engineering: overview, scope and challenges. Biotechnol Appl Biochem 39:29–47
van de Pavert SA, Olivier BJ, Goverse G, Vondenhoff MF, Greuter M, Beke P, Kusser K, Höpken UE, Lipp M, Niederreither K, Blomhoff R, Sitnik K, Agace WW, Randall TD, de Jonge WJ, Mebius RE (2009) Chemokine CXCL13 is essential for lymph node initiation and is induced by retinoic acid and neuronal stimulation. Nat Immunol 10(11):1193–1199
Vozzi G, Previti A, De Rossi D, Ahluwalia A (2002) Microsyringe-based deposition of two-dimensional and three-dimensional polymer scaffolds with a well-defined geometry for application to tissue engineering. Tissue Eng 8:1089–1098

MIX
Papier aus verantwortungsvollen Quellen
Paper from responsible sources
FSC® C105338

If you have any concerns about our products,
you can contact us on
ProductSafety@springernature.com

In case Publisher is established outside the EU,
the EU authorized representative is:
Springer Nature Customer Service Center GmbH
Europaplatz 3, 69115 Heidelberg, Germany

Printed by Libri Plureos GmbH
in Hamburg, Germany